CATTLE

AND

The Future of Beef-Production
in England

CATTLE

AND

The Future of Beef-Production in England

BY

K. J. J. MACKENZIE, M.A.

READER IN AGRICULTURE IN THE UNIVERSITY OF CAMBRIDGE,
LATE EDITOR OF THE JOURNAL OF THE ROYAL AGRICULTURAL
SOCIETY OF ENGLAND

WITH A PREFACE AND CHAPTER

BY

F. H. A. MARSHALL, Sc.D.

FELLOW AND TUTOR OF CHRIST'S COLLEGE AND UNIVERSITY
LECTURER IN AGRICULTURAL PHYSIOLOGY, CAMBRIDGE

CAMBRIDGE
AT THE UNIVERSITY PRESS
1919

CAMBRIDGE
UNIVERSITY PRESS

University Printing House, Cambridge CB2 8BS, United Kingdom

Cambridge University Press is part of the University of Cambridge.

It furthers the University's mission by disseminating knowledge in the pursuit of
education, learning and research at the highest international levels of excellence.

www.cambridge.org
Information on this title: www.cambridge.org/9781316605509

First published 1919
First paperback edition 2016

A catalogue record for this publication is available from the British Library

ISBN 978-1-316-60550-9 Paperback

PREFACE

IT is now more than thirteen years since Mr Walter Heape in a work on "The Breeding Industry" pleaded for "the establishment of a State Department of Animal Industry, organized and controlled by a staff fitted by their training; first, systematically to record the condition of the industry throughout the Kingdom; secondly, to deal with the many problems breeders from time to time require solved; and, thirdly, to present the result of such work to breeders in a practical form." Opinions may differ as to how such work may most profitably be carried out, but that the need for it, at any rate as regards cattle-breeding and beef-production, is as great as, or greater than, it ever was, must be clear to anyone who reads the present work to which these remarks are by way of preface.

It may be objected that the breeding of domestic animals is an art and not a science, and that the great breeders of past generations, like all successful judges of live-stock, were men highly endowed with those special attributes of hand and eye which together compose the peculiar faculty of selecting, for the perpetuation of their kind, beasts of a type most serviceable to man. There is much truth in this criticism. In choosing an animal at a sale or judging it at a show, it is usually impracticable to adopt methods of precision; at any rate these are seldom or never employed, for the judgment is arrived at after a survey which though often hasty may nevertheless be wonderfully correct. Yet it is obvious that the art of selection, valuable as it is, would be of incomparably greater service if supplemented by accurate anatomical knowledge and an appreciation of the varying degrees of functional value possessed by the different parts of the animal. As Mr Heape expressed it in the work already quoted "the condition of the breeding industry now is no whit better than was the art of sculpture before anatomy became a science, and by the application of science the breeder will gain no whit less than the sculptor gained by a knowledge of anatomy."

The scientific breeder, however, requires something very much more than a mere general knowledge of the form and structure of an animal, and unfortunately for the progress of this important branch of study, the necessary knowledge at present hardly exists. It is, of course, true that the points or characters which contribute towards making a good beef-beast or a good milker and the points of all the more important breeds of live-stock are definitely laid down in the text books or in official works of reference. Yet of the anatomical and physiological factors which go to constitute these points we know little or nothing. And we do not know with any degree of accuracy how many of these points have an inherent value in virtue of possessing properties of direct economic importance, as with a good shoulder, a good loin, or a well-developed leg of mutton. We are still more ignorant concerning those points which derive their value, whether real or imaginary, from being associated with other characters which are of recognized importance but are obscured to the superficial observer; the shape of the head in the improved Shorthorn is an example of this kind. And again there are almost certainly some points the value of which is wholly fictitious, or at the best highly problematical; these are the "fancy points" which depend simply upon caprice or fashion, such as the white face of the Hereford cattle, or the ruby-red coat colour of the Shorthorn, which for no palpable reason is so often preferred by expert judges to the brick-red colour. Where so little is known the field of investigation is almost unlimited. For before scientific breeding can become an assured success, it will be necessary, first, to determine accurately the characters of live-stock which are of economic importance (whether directly or by reason of their possessing a high degree of correlation with other points which are of value), and secondly, to investigate the relative degree of utility of these characters. It will next be necessary to ascertain the anatomical composition of these points, and their functional or physiological relations, and not till this has been accomplished will it be possible to build on sure and firm foundations, and to make a profitable study of the laws which govern the inheritance of the various points both individually and in combination.

The truth of these propositions is abundantly illustrated in the pages of this book, in which the author has set forth the results of his long and varied experience as a practical farmer and as an investigator and teacher of scientific agriculture. The importance of Chemistry and Botany to agricultural practice has long been recognized; the application of physiological science, which hitherto has been almost wholly limited to the domain of nutrition, is equally essential to a sound animal husbandry. If this country is to maintain its present leading position as a producer of high-class live-stock and at the same time to keep up a due supply of meat and milk for national consumption, it will be necessary to utilize every possible advantage that can be derived from the study of biological science.

F. H. A. MARSHALL,

March, 1919

AUTHOR'S PREFACE

I AM aware that this book deals with subjects that are very complicated, and that no one man can have an extensive knowledge, born of personal experience, of all of them. It is therefore a matter of great regret to me to know that parts of the work will cause consternation to many personal friends and acquaintances, to whose help I owe the larger proportion of such knowledge as I possess. I know only too well that much of my text will strike a blow at these practical men in that which, after their honour, they cherish most—their prejudice. I have therefore to ask my friends and others who are of the cattle world to remember that I have not hurriedly taken up my pen to write this, my first book. It is over thirty-six years since I had my first lesson in the selection of stock from an experienced, successful and enthusiastic breeder. During the past twenty years of my life it has been my duty, as well as my pleasure, to pick the brains of all the fraternity, from the drover to the pedigree expert, whenever I have had the good fortune to meet one kind enough to help me, knowingly or unknowingly, in my search for knowledge. Further, I would ask all those who feel aggrieved to realize that I have had to banish the same prejudices from my own mind, which was once as full of them as any British husbandman could wish it to be. I can only hope that anyone whom I may unfortunately offend will try to imagine the long and strenuous struggle which such a change in my state of mind has cost, and so forgive, even if he cannot approve, this record of the change in my opinions which a long search after truth has brought about.

I have to thank my colleague, Mr R. H. Rastall, for reading the slip-proofs of my book. I cannot, as they are so numerous, thank by name all the practical men for the help I have just alluded to, so I must ask them to take the will for the deed. I must, however, express my deep sense of gratitude to my colleague, Dr F. H. A. Marshall. He has read and revised the proofs, and greatly added to the value of my first effort at book-production by writing a preface and the chapter on Physiology. Further, during the last ten years, the whole of which time we have worked in unison, he has not thought it beneath his dignity as a man of science to come to my aid when I was seeking knowledge likely to be useful to the farmer, the butcher and the rising generation of young agriculturists, as well as to the consumer. For his large-minded readiness to investigate subjects intimately connected with the pig-sty, the byre and the slaughter house, I (and I venture to say all practical men) owe him very many thanks. In trying to express my gratitude to Dr Marshall I would like to add the hope that his example will be followed in the future by many of those eminent scientists who now regard the problems of animal husbandry as unworthy of their notice. Even the great kindness Dr Marshall has shown to my poor effort cannot be mis-spent if it leads to other biologists taking up, and working out systematically, some of the many problems now demanding solution in the life-history of the animals that supply food and clothing for the human race.

K. J. J. M.

March, 1919

CONTENTS

CHAPTER I

INTRODUCTION

EVEN before the war there were some who realized the difference between the process of "stealing from the land" and the
operations of farming; and among this small minority there
were many who saw that land kept under permanent grass was
more suitable material for the thief than for the honest producer.

But since August, 1914, very many—perhaps the majority—
have come to realize that their comfort in life is, to a very great
extent, more dependent upon food than upon luxury; and that,
without farming, the produce of this island-home of ours is not
sufficient to keep the inhabitants decently fed even for a fairly
large part of the year. So from both sides there has lately been
a clamour for the plough; it has been maintained, quite rightly,
that fields which are worked deeply, manured skilfully, and
seeded properly are likely to yield food in greater abundance
than land left to cover itself with a herbage whose quality varies
with the natural fertility of the soil and with the bountifulness
of our uncertain seasons. Further, some of the majority are now
inclined to join a small section of the minority who never tired
of insisting that, if the British farmer would but make an imaginary
journey across the Channel or the North Sea and emulate the
agriculturists of Eastern and Central Europe, many difficulties
of his situation would vanish.

That those who insisted upon the good that might come of
a study of the arable husbandry of Denmark, Holland, Belgium
and Germany had much reason upon their side is obvious to
all who have investigated the subject of food-production; but
the fact that the conditions which favoured success on the continent were widely different from our own was not sufficiently
kept in view. Many enthusiasts, indeed, spoilt a good case by
exaggerating it, but their contentions, though somewhat extravagant, were especially valuable when expressed in the worst

period of public apathy, and deserved at least to be treated with thoughtful criticism.

One of the subjects that seems to have been overlooked is that of beef production, and it is the object of this work to show how the continental practice must be very considerably varied if we are to maintain our supply of the "Roast Beef of Old England."

I once happened to be waiting as an expert witness in court when one of His Majesty's Justices gave a short dissertation on the nomenclature of various articles of food. He explained that to the expert there were differences of terminology which might be of subtle or of emphatic degree: "For instance," said his Lordship, "to the grocer there are new-laid eggs, fresh eggs, *and eggs*!" Now, without presumption, I hope, I would follow his Lordship's example, and point out that beef, to the Englishman, is quite different from that grown on the plough-lands of the continent. We have to recognize this factor more fully before we are in a position to reorganize our husbandry.

Let us for a moment review the cattle husbandry of the continent. Obviously, to do this briefly, one must generalize. To review the subject in detail would demand a very much larger volume than the present, but we must at least attempt to visualize our neighbours' conditions if we are to measure home conditions by their standard.

Our neighbours use their cattle primarily with a view to the making of butter and cheese, to supply the milk-salesmen, and for draught purposes. Meat, though important, is quite secondary. Their cattle supply meat in the form of veal, cow-beef, and ox-beef; and also, strange though it may seem, as pig-meat. Whey and separated milk, the by-products of their most important industry, the dairy, are the means of manufacturing very large quantities of bacon and pork. During the war our farmers were urged to graze their pigs on our permanent grass-land—wise counsel for Englishmen no doubt at the moment, but a measure that would be looked upon as the strangest extravagance by continental farmers, who regard the pig as most valuable when used to consume stuff that cannot be used more profitably for anything else. When they keep land under permanent grass, which is not often the case, it is simply with a view to the cow directly producing human food, such as butter, cheese, new milk and veal; waste material is, in their view, quite good enough for the pig. Agricultural conditions which allow the products of the soil to be consumed by animals which, after a period of slow growth, will appear upon the table as meat, hardly enter into the continental view of farming. With us, on the other hand, beef is a most important product, perhaps the most important, as regards cattle, of all our grass-land products; and, before the war, much more than 50 per cent. of our land was permanently kept under grass. The only product of our cattle-husbandry to be compared with beef was the new milk, required by large residential districts; for our cheese-making industry, whilst still important, was, and is, very small, and our butter-making, except as an adjunct to calf-rearing, had disappeared in the majority even of the most rural districts. Certain parts of Ireland must be excepted from this last statement, but even in the case of Irish farmers, it is doubtful whether calf-rearing or butter-making is the more important. Amongst English farmers it is a common practice to devote about three acres of medium quality grass-land to their cows; in return they get, per annum, one well-reared calf and slightly increased bulk in the cow. It is this kind of pastoral husbandry, forced upon us by the economic conditions prevailing since about 1875, that is in people's minds when they urge pig-grazing upon the notice of our grass-land farmers. The Dutch farmer might well be amazed at the idea of using some of his magnificent Polder pastures for pork production; he only knows of this land as being used for growing milk. An acre of his land will yield him approximately 300 gallons of milk, whereas our very best grass-land does well if it produces 280 lb. of prime bullock, equivalent to 160 lb. of meat. Though it may be possible to show that pigs fed upon grass-land will produce more pork than the bullock will produce beef, it cannot be claimed that, under the most favourable conditions, they will produce the same amount of human food as the milch cow.

Here and there the foreigner does grow some prime beef; it is not an unknown thing in his husbandry, but the process is so seldom favourable to intensive production of food from the land that the practice is almost negligible. The beef which the continental *chef* knows how to serve so well is practically all cow-beef, with an occasional joint off a good young bull, or meat from draught oxen, the oxen being fattened for slaughter after a long life at the yoke. Mutton, as travellers know, is a rare luxury on the table of any European country except our own. Even with sheep the foreigner is not negligent of the dairy, for much of the mutton he eats is the flesh of animals that have been milked. Ewes' milk in our country is practically unknown for any other purpose than the rearing of the lamb. It is quite amusing to see the astonishment shown by some foreign agriculturists on hearing that we never milk sheep, and the corresponding amazement displayed by our farmers on hearing of such an anomaly.

But while the foreigner consents practically to abstain from good steer beef, he is not a little careful that his cow and bull beef is of uniform, and of fairly good, quality. The huge cow-market at Leeuwarden in the Friesland province of Holland is a wonderful example of this. At the great cow-market held at this great agricultural centre are to be seen vast quantities of fine cows ready for slaughter. What strikes the Englishman about the market when he visits it for the first time, is the wonderful uniformity of the stock; row upon row, each containing several dozen specimens of the cows of the country, are all more or less exactly turned to the same pattern. The cattle are all of the same type, not particularly good (the best cows in an English market are undoubtedly better), but there is practically never a bad one. The cows are all fairly young, being from seven to nine years old; a wastefully fat animal is never seen, and they are practically all in the same stage of "finish," —what would be called "just good meat" in our home markets.

The bulls are remarkable to us in one particular respect. Practically all of them are about 30 months of age or a year younger. An old bull is an exception, being just the odd one who, by virtue of his breeding and appearance, has been selected by one of the Associations for the Improvement of Cattle as worth subsidizing; thus his services as a sire remain available for four or five or even more years, the less perfect bull being slaughtered after one, or at most two, years' service.

On one visit to Leeuwarden, I remember seeing, at the abattoir belonging to a large firm of exporters, no less than six hundred "sides" from the carcases of young bulls that had been purchased on the market that same morning for shipment to London by one firm; and yet that very morning I had searched in vain for a veteran. For, of course, these old bulls are the choicest specimens of their race, and I was anxious to inspect any that happened to be on the market.

If one contrasts a market full of animals such as this with the good and the bad, the old and the young, the lean and the extravagantly fat found in any of our large sale-yards, one does not wonder at many of one's countrymen not knowing that cows and bulls yield wholesome food. On one occasion a young Englishman, of quite average intelligence and well-informed in many matters, asked me if I was allowed by law to sell my cows, once I had done milking them, for human food! I am not at all sure that he believed me when I told him that 80 per cent. of the beef he ate when travelling on the continent was the flesh of such animals, and I fancy that most of our English tourists are much in the same state of mind as my friend.

This class of meat must, however, be held to be inferior to our prime joints, and though by good cooking the cow-beef of the continent may be brought very much nearer to the prime "Roast Beef of Old England," it will always be its inferior. To imagine ourselves a nation of cooks is difficult, but it is easier to do this than to imagine that our national standard of living should fall to the level of cow-beef served from the kitchen of the housewife who for the past generation has had nothing in her larder but good English meat. For myself, I am content to hope for the day when the average English cook will seldom, or never, spoil the prime article which the profligate state of our pre-war supplies had made super-abundant. If the continental meat supply were to be forced on our people as an immediate consequence of the war, their sufferings would be considerable, for it would take a generation, at least, to train a class of cooks that could be trusted to send it to table in a palatable form

Even then one has to assume that the Englishwoman could be persuaded, and trained, to take the infinite pains with the details of her household work that such cooking demands. One is inclined to doubt whether she has the instinct and capacity for such work.

In the absence of first-class cookery or prime meat, it is as easy, as it is unpleasant, to foresee a general fall in our national standard of life—at any rate as regards the food which serves to prolong our lives. Nobody with any knowledge of the well-paid labour class, will for a minute believe, if he has imagination enough to realize what such a change in the national food would mean, that our countrymen would tolerate such a condition of affairs under any circumstances but those of dire necessity. There is no doubt that, should the necessity unhappily arise, such a change would have a very pernicious effect on the efficiency of our race. The well-fed man is a contented man and *vice versa*, and the more contented a man is, the more likely is he to be the head of a useful family—and the State is but the reflection of the family. Yet this change, with all its consequences for evil, is the one that is urged upon our life by those who would import continental methods of agriculture to replace our own *in toto*.

These authorities are at one with all who are patriotic enough to deplore a return to the state of affairs prevailing in August 1914. The British public of those days dimly, if at all, realized that agriculture was connected with the people's food. The U-boats, it is true, have taught us how dangerous it is to be dependent upon transport for food. The folly of relying upon lands on the other side of an ocean, while a large part of our own land was unproductive, has been demonstrated only too well. The rationing forced upon us by U-boats has done more, in a few months, to make people think of the fruits of their own land than writing, platform oratory, or argument had done in decades. Nevertheless, it is the height of folly to expect that the public will altogether forgo the best type of food as a result of such lessons. The ordinary man will get the best, particularly when he has been brought up to expect it, from overseas, if it cannot be produced for him at home. There is still plenty of land left on the earth's surface from which to rob, and while this is so, our own good breeds of beef cattle, exported in considerable numbers for the purpose, may be relied on to convert the produce of the untilled plains into prime meat. At a price, the carcases of the descendants of our much-boasted pedigree stock will be returned to us in admirable condition in the freezers of ocean-going ships. Though the amount of soil awaiting the land-robber is limited, there is enough of it left to last even until Europe has recovered from this war. The dread of the next resort to arms will not be enough to prevent our people from sending money across the ocean in return for the produce of such lands for so long as the supply lasts. A nation accustomed to prime meat is more than likely to go on eating it while it can, even though it can be shown that its place of origin is insecure. That the people will have their meat in peace-time, whatever the cost, unless and until their patriotism is awakened by their country's danger, seems to be the only assumption upon which an agriculturalist who is making plans for the future may work. This assumption demands that any reform of farming practice must combine intensive farming with the most economical production of prime beef.

The necessity of reform is obvious, if safety is to be considered worth attaining. The United Kingdom has, it will be admitted, not made herself safe from the tyranny of evil-minded and rival foreign countries in the past; she has left the satisfying of her people's hunger to others, she has had no care for the produce of the land which has been entrusted to her. She was, in August 1914, as vulnerable to starvation as any uncivilized country; she has to thank the indomitable spirit of her people that her lack of foresight did not lead to her destruction for want of the necessities of life. She ran the risk so that she might boast of her food being cheap, so cheap that her people learnt to waste that which they have at last learnt, after forty years of profligacy, to value at something approaching its worth. That she does not wish to return to the unhappy conditions prevailing from 1875 to 1914 may be assumed; yet she has the right to demand that her foodstuffs should be as far as possible produced from her own soil, that the foundation of all life should be produced from the land of these islands with all the intensity and reasonable economy that good brains, sound training and hard work can supply. To do this involves some change in our cattle husbandry, but it involves, further, a reformation in the whole of our agricultural community.

The landlord must realize that to lead his own class intellectually is his first duty, and that failure to do that work, whatever may be his services to the rest of the community, leaves his first and greatest obligation undischarged. The landlord's clients—the farmers—have to forget their prejudices and learn that their profit alone is not all that is asked of them in their work for the State. They have to realize that the unfortunate past is to be forgotten, and that the future demands that they combine *production with profit*. The farmer's colleagues—their labourers—have still more to forget. They must learn that in the future the State does not call for underpaid drudgery unwillingly given, but means to have intelligent labour willingly given for a living wage.

CHAPTER II

STORE CATTLE

THE production of lean animals for the feeder to fatten is the work of the commercial breeder of horned stock. Sometimes the complete process of breeding, rearing, and fattening is carried on by the same person, but it is exceptional to find animals offered to the butcher that have spent their whole life as the property of one farmer. The history of an animal wanted for beef is generally much more varied. It is born on one farm and perhaps weaned there, though very many move during quite early infancy; then, after the period during which it is known as a calf, it becomes a "store," and, as such, often has several homes in various parts of England. The importance of the store-stock trade is characteristic of our agriculture, and there is certainly no other country of Europe in which this class of live stock plays so prominent a part as it does with us. The large amount of traffic in this type of cattle is due almost entirely to the change which bad times brought about in our husbandry some forty years ago. Before then less than half of our land was under grass, the larger part being under corn and root and other fodder crops. The proportion under permanent, as distinguished from "temporary" or "rotation," grass was confined very largely to two classes of soil. We had, first of all, the land which was so good in itself that it produced very well without the necessity of working, manuring and seeding it. Though there were not many such fields, they were enough to play an important part in our husbandry. Their produce, when well managed, was sufficient to justify the best husbandmen leaving them unploughed. Indeed it was only when corn should reach an indefensible figure, that it could be hoped to obtain any reward for the enterprise of breaking them up. If their produce while under grass, as pork or as milk or as beef, had been averaged, it would have yielded approximately the equivalent of 1500 lb. of grain, while under the plough possibly an additional thousand pounds of cereals might have been obtained every other year. This increase, even allowing for the yield of straw, was not likely to be of enough value, nor was it desirable that it should be so costly as to pay for the outlay on horse and manual labour, manure and seed—to say nothing of the interest upon the extra building accommodation usually wanted for such cultivation. The very best of this land, which was supplied by nature with watering places, yielded the primest beef, or, in exceptional cases, mutton; acres not quite so perfect fed milk-giving cows; another class, generally because it was not watered, supplied hay for the wintering of farm stock and also for the large numbers of horses wanted for industrial purposes in our large cities. Land of the highest natural fertility was, then, one of the two classes of soil left unmoved by tillage implements.

Let us now consider the other class. This second class of land was left uncultivated because it did not pay, even when prices for agricultural produce were good, to move it with implements of tillage. It might be that the land was inaccessible, that it was not of such a nature as to yield plant food; it might be too dry or too wet—but, for one reason or another, it did not pay to work it. Prices of produce, which must always fluctuate to a greater or less extent, obviously make the degree of worth-lessness, which constitutes uncultivable conditions, a changeable factor.

Such land (and even the most worthless yields some produce), since it would not return anything to the good farmer who tilled it assiduously and with skill, was the *justifiable* prey of the land-robber; and husbandmen, good or bad, will always continue to steal from it.

Between these two classes of land, the best and the worst, lies the greater part of our food-producing soil; the fields which will yield abundantly when well worked and manured, but lack the inherent fertility to produce largely when unculti-vated. Even of this land there was always a certain proportion under permanent grass; a small, but appreciable, proportion of grass, of course varying in extent with circumstances, has always been, and is always likely to be, found on most English farms. Our climate is so changeable, the formation of our country is so crumpled and diverse, that more often than not it is found wise not to have too many eggs in one basket; and so it will nearly always be found that a good homeland husbandry will justify a moderate proportion of permanent grass on the farms in districts where the soil favours an arable holding—even at times when cereals are selling at moderately high prices. This is probably the explanation of such a large proportion of grass being found in England in the days before corn-growing became a ruinous undertaking on most of our farms.

Without wearying ourselves with the innumerable dates and figures, it may be said that the store-stock trade became paramount in our agricultural economy for four reasons:

(1) The continuous fall in the prices of cereals which led to the "stop the plough" policy[1].

(2) The continuous downhill grade of values for mutton and wool in the last quarter of the last century which led to a great reduction in our flocks of sheep. The number of sheep in the United Kingdom was 32,246,000 in 1872 but had fallen to 27,629,000 by 1913.

(3) The recurrence of outbreaks of contagious disease among cattle, caused, as it was held, by imported store and other cattle. These animals of store-stock grade came to us from the vast wild plains of the New Worlds overseas; their bodies were, frankly, the produce of simple robbery of the herbage on virgin soil. Whether these cattle did, or did not, import disease is a matter of controversy, but the point need not be argued; for, as long as they were freely imported, it was found impossible to control disease, and so, after many temporary periods of exclusion, they were finally excluded for good and all by the Diseases of Animals Act of 1896.

(4) This exclusion led to the fourth and final reason for the great development of the store-stock industry, though it is really involved in the other three.

We have seen that the area under the plough had greatly

---

[1] The arable area in England fell from 13,800,000 acres in 1872 to 10,800,000 in 1913. But this does not exhaust the loss of bread-growing land; a *smaller proportion* of the area still under tillage was used for growing wheat in 1913 than was the case in 1872.

decreased, and that the number of sheep to graze on the increased area of grass had declined. Why was their place taken by store cattle? Chiefly because land could only be kept under corn when and where it was possible to produce very high yields of cereals, or to combine corn-production with some more remunerative crop like the potato. Such crops demand very high manuring, and our farmers came to rely upon large doses of rich farmyard manure with which to improve their land and so make it produce enough to pay for ploughing, even when wheat stood between 25*s*. and 30*s*. per quarter of 504 lb. weight. Lean or store cattle of age enough to stand very large rations of concentrated food and big enough to tread underfoot great quantities of litter while consuming incredible quantities of roots, were an admirable means of making the richest possible "dung" in sufficient bulk to satisfy the demands of the land. While the store cattle from overseas were coming in large numbers, the farmers could command a supply at reasonable prices. That is to say, they could buy lean stock at a cost which allowed of the cake, corn, roots, and hay being paid for out of the increase in price which the fat animal made when sold for the butcher; the straw used as litter was usually thrown in as a product that was useless for anything but farmyard manure-making.

When the supply of imported stores was stopped, the corn growers had to fall back upon the home-bred supply. Buying on a market that was not an open one, the feeders of beef often created a demand that was greater than the supply. Thus even in the most unhappy days for farming, the growing of store cattle was, to a certain extent, remunerative, while feeding became a very extravagant process of manure-making. Frequently the difference between the buying-in price of the lean bullock and the selling-out price of the fat animal hardly paid for the cake and corn, which was only too often given to the beef beast in inordinately large quantities; the hay and roots were left uncashed, and the cost of their production absorbed in the expenses of manure-making. This kind of farming, it must be realized, involved two, or at most three, acres of corn carrying the charges on an acre of roots, as well as the expenses of their own growing.

I am not concerned to defend such a practice. On the contrary, I agree with the many, though not the majority of, agricultural authorities who have condemned this form of waste-fulness. For over ten years I have spoken and written against the useless extravagance of supplying plant-food through the excreta of overfed bullocks, and I am thankful to know that I have not worked altogether in vain, though the pernicious system was still very prevalent when war broke out.

It must, however, be remembered that there was much excuse for this extravagance. It had behind it a very long tradition, and tradition is always strong in the complicated business of farming; experience, moreover, had shown rich farmyard manure to be not only the best, but almost the sole, means of supplying the land with a full amount of available plant-food. This was indeed the case before the advent of the agricultural chemist. "Good Horn Good Corn" was the motto of every practical man whose empirical knowledge had been handed down to him through several generations of successful farmer-ancestors.

At first the chemist did little to improve the ordinary plan; and the process of learning nature's method of contriving the marvellous food-manufacture in the soil is so complicated, that it may almost be said that for some time the man of science did much to confirm the practical man in the belief that "chemical" manures could not replace the plant-food in real good "muck."

But Lawes and Gilbert through their research and field-trial work at Rothamsted demonstrated that plant-food had many other sources than that which came from rich cakes passed through animals to the manure-heap. The great work of these masters was, however, much handicapped by some of their followers, who went beyond their teaching in unduly pressing the claims of the stuffs that could be carried about in a sack. These early followers, in their enthusiasm for the valuable con-stituents of the concentrated fertilizers, were apt to forget the value of humus contained in farmyard manure, and, further, to overlook the prime necessity of using humus and implements to obtain proper texture. They were liable, too, to omit from their teaching the fact that weeds, when not mastered by good sound tillage, throve on the plant-food from artificials at the expense of farm-crops. In spite of the fallacy of much of their propaganda, these men did good work; but when others, profiting by their mistakes, taught sound theory, the State was very slow to supply means to demonstrate and spread their teaching.

This negligence, though it was as potent as it was lamentable, was not the chief cause of many plough-land farmers continuing to buy store cattle at too high a figure and to overfeed them while fattening them. The most pernicious influence of all was, in my view, the bad example set by those who ought to have been the leaders in improvement. The shows held by Agricultural Societies for live stock encouraged the exhibition of extravagantly fat animals, and, in the case of breeding stock, of disgustingly overfed creatures. The landlords were prominent exhibitors at such gatherings. These breeding animals were not, it is true, overfed for purposes of encouraging the tenant farmer, but as a means of securing high prices which the foreigner would pay when an animal, otherwise a perfect specimen, was really "well up." The condition of being "well up" may, I think, be exemplified by an account of a conversation I had with one of the greatest veterinary pathologists in Europe.

" *The condition of fatness,*" he said to me, " *to which you get your show cattle is undoubtedly pathological.*" This hideous state of fatness, where breeding stock were on exhibition, was almost greater than at our fat-stock shows. This may appear, at first sight, incredible, but there is an explanation.

The butchers were so tired of all the dripping, tallow and lard which the carcases of the exhibits carried, that they set their faces against the prize-money being given to specimens that were altogether too obese, and their influence has been sufficient to moderate the excess to a certain extent. At the breeding-stock show, on the other hand, the foreign buyer seems only capable of judging a first-rate specimen when the whole body is covered with an excessive coating of firm grease.

That the landlords of this country should have taken a leading part in these contests was regrettable, but the harm done by them as exhibitors did not end the trouble.

I know most of the fat-stock markets in England and many

of those in Scotland, Ireland and Wales, and I have seldom been at one without seeing overfed specimens straight from the "home-farm." As the war has once again proved that this over-feeding is opposed to a really efficient national agriculture, it may be assumed that these wasteful and extravagant customs on the part of land-agents will be stopped, and that the land-lords of the future, looking into matters themselves, will insist upon a more scientific practice and one less in sympathy with methods which involve wholesale theft from the land.

The rather uneconomic demand upon the supply of stores from the arable farms re-acted upon the graziers who wanted to use their rich grass-land to finish stores. It must be remembered that this so-called "finishing" land became more plentiful as bad times became worse.

Many farms had plough-land fields good enough for any purpose. The owners, or tenants of such fields, were under great temptation to lay them down permanently to grass. The owner was assured of a higher rent if he did so, grass-land property demanded less outlay upon buildings, and it was easier to find good tenants for such land. The tenant had less difficulty with labour, he was able to pay a better wage, he was less dependent upon the vagaries of our uncertain climate, the heavy local taxation on buildings as they fell into disrepair and were finally abandoned had not to be met, and, altogether, grazing land was a less risky and much less troublesome business than arable land farming. No one can complain of the transaction under the conditions which existed when the change was brought about, but inasmuch as store-growing demanded to a large extent a system of land-robbery, anything that increased the profit of the venture was to be deprecated from the point of view of the home production of food.

That the store-cattle industry, as generally carried on, was robbery from the land, a study of the life-history of one or more animals of any class will show, and examples will now be given.

Store-bullock reared on pasture.

We may begin by taking an animal bred for beef from non-pedigree parents of one or other of our several large beef breeds, and, without having been through any period of privation, finished as prime meat on good Midland bullock pasture at the age of three years and six months, or just when his dentition was nearly complete. We will assume that this animal was born in the month of May, that he found himself in the first moment of active existence, say, at 48 hours old, running beside his mother in a district of well-shaded and well-watered store-grass land in a county of humid climate. He thus began life among rural scenery remote from the cities of men, but as beautiful as any in the world. Running by his mother's side his first six months of life were happy; his food was plentiful and of the sweetest, new milk and soft young grass in abundance being his daily fare, and, for amusement, the constant companionship of other youngsters of his own kind and age made life perfectly happy except for the attention of his natural enemy, "the fly." This pest is important to all concerned in his welfare; for much of the energy of the little calf's food is wasted in his frantic, if foolish, gallops to escape these tormentors. The insect pests of the order of the Diptera play a very pernicious part in the summer-time growth of the store throughout his life and greatly reduce his profit-giving capacity. At the age of about six months the calf was weaned, and, supposing he was destined not to travel for another year, his first winter was spent in the open yard and shed attached to the farm on which he had been born. This, his second period of life, was likely to be less pleasant; his food consisted of the grass he could pick up in open weather and an ample, possibly wasteful, supply of hay of moderate quality with cold water to drink. His first birthday found him under happier conditions, for the supplies of young spring grass reminded him of his first youth and provided him with nourishment and pleased his palate till about his eighteenth month. Being "six quarter" old he was sold, as we will suppose, by his breeder, and for the first time knew the troubles of the young bullock in the market-place, and felt the inconvenience of travel in the cattle-truck.

For the sake of simplicity we will suppose that the young beast is sent by rail to some county where it is possible to winter him out on aftermath grass or on medium pasture that has not been too heavily stocked in the summer or early autumn months. On this grass he is given enough hay to "keep him from going back," which produces a slight increase in growth. Another summer's pasturage is followed by a winter's keep similar to the one last described, except that, being older, he is made to depend more upon pickings off the land for his living; his supply of hay is probably the same; the increased size of his frame uses up more food for its sustenance and allows of a smaller residue for growth; consequently, such a store-bullock will put on very little weight during this, the third, winter of his life. Being now what is described as a "Six tooth" he is turned into a field of finishing pasture to wax and grow fat and to furnish, after some 5 months' grazing, a prime carcase of grass beef.

Having briefly sketched the young bullock's imaginary career from the calving-box to the block, we may tabulate its stages to show more precisely how the produce of the land was used and what return was made to the country. It will be noticed that in this example no food other than grass and hay, either through the mother's milk or the animal's own mouth, has been consumed. This is a class of beast much valued by the grass-beef maker, for he dislikes those that have been pampered in yards with cake, roots and other dainties. In the following table the finishing process, which will be included in another chapter, is omitted:

Life-history of a store-bullock bred and reared on pasture.

Situation	Age	Amount of land required	Weight at end of period
On a breeding farm Medium "store-land"	Birth to 6 months	2·5 acres[1]	400 lb.
,, ,,	7–12 ,,	·3 ,,	500 ,,
,, ,,	13–16 ,,	·60 ,,	700 ,,
On store rearing farm	17–24 ,,	·70 ,,	800 ,,
Good "store-land"	24–30 ,,	1·0 ,,	980 ,,
,, ,,	31–36 ,,	0·75 ,,	1050 ,,

Result at 3 years from 5·85 acres of land: 9 cwt. 1 qr. 1 stone of beast.

[1] Includes keep of cow for 12 months after deducting the amount, about 15 %, that one might expect to be used up in causing the growth of her own frame if she were managed with skill.

M. 2

From this general statement which, taking good seasons with bad, the best growing animals with the poor doers, and the most skilful and careful management with the most foolish and slothful, is a sufficiently accurate guide, several conclusions may be drawn:

First, that nearly 6 acres of land yield 1050 lb. of live beast or 180 lb. per acre[1], which, assuming a carcase-yield of 54 per cent., represents 567 lb. of side of beef plus a hide weighing round about 70 lb. and some useful offal. Reducing this all to meat and bone value, we may say 700 lb. of useful stuff—about 120 lb. per acre, a figure which demonstrates at once the paucity of output, from the national point of view.

However, as it was produce obtained from the land that demanded practically nothing in return, it was often the only system in England showing a profit. For the big outlay in the transaction was rent, and during the last 30 years it has always been found that the landlord can be squeezed; in fact, it is only too true that during this period many wealthy men became landed proprietors in order to enjoy the amenities of being in a position to be squeezed.

The practice was also pernicious from the business point of view, since the turnover of money was very low. A three-year-old store was well sold for £22 and very cheap at £16 in the days of peace; this gave a turnover of but £3 to £4 per acre per annum.

On the other hand, the system allowed of better wages being paid to the individual worker on store-raising than on plough-land farms. One pound a week, with cottage and garden, could be earned on a store-raising farm as against about 17s. given to the labourers on arable land and, though the work involved long hours, it was more exhilarating and less drudge-like. It bred a fine type of man.

Store reared on plough-land and grass.

The store required for the making of farmyard manure by the feeders in the Eastern counties of England and Scotland

---

[1] I have allowed for a certain proportion, say 1 in 4, of heifers which would not be so weighty as steers.

would probably be slaughtered at the age of 35 months. That is to say, the steer or heifer would have been bought for fattening when just two years and a half old.

To obtain a complete picture, many stores from the grasslands (going for the same purpose at the same age, but more often having much more varied careers) would need to be selected. The calf would have been dropped by a cow kept for milking in mid-autumn, transferred immediately to some district where butter was made, and the skim, or separated milk, given to the young animal together with some artificial food to replace the cream—some hay, a few roots, etc. A paddock would find him nourishment as soon as spring grass was available, and his liberty would be curtailed through a long winter in a straw-yard, during which time he would receive a very small proportion of corn. Grazing on grass of second quality in spring and summer, and protection while feeding on straw and other plough-land produce in the homestead in autumn and winter, would continue to the end of the steer's store career.

The following statement attempts to give an account of how such a steer uses up the area of land required to rear him and to show what return may be expected for the enterprise. We will suppose that the calf is born early in September, and goes into the feeding-courts in October, when just over two years old. To make the example simpler we will assume that only home-grown corn is used; as regards rough fodder we will also assume that the straw grown with the corn is fed. A slight economy might be shown in pounds, shillings and pence by using imported feeding stuffs, and some linseed would make better feeding. So the increase in weight shown will be as great as if the best food, in proportional amounts, had been fed. For instance, rather less linseed cake would give the same nourishment and growth as the 2 lb. of oats allowed while the animal is being reared on the pail. In the same way we will take hay, equivalent in weight to oat straw, when a less palatable and nourishing article of diet would suffice. As regards the increase in weight, allowance is made for the many ills to which calves reared on this system are liable. Such animals seldom recover from their bad start in life without much additional and very costly cake-feeding. The following is an estimate only; but, like the last example given, it represents what I have many times seen in a fairly long experience.

Store reared on plough-land produce and on grass.

Foods fed	Age	Land required	Weights
Milk, 50 gals. Separated milk, 100 gals. Oats, 300 lb. Oat straw, 400 lb. (for cow) Hay {fed to calf, 150 lb. {for cow's milk, 500 lb. Roots, 300 lb.	Birth till end of 5th month	Acres 0·145 of oats ,, 0·2 of hay ,, 0·005 roots ——— 0·350	300 lb.[1]
Oats, 300 lb. Hay, 400 lb. Roots, 500 lb.	6th to end of 8th month	Acres 0·145 of oats ,, 0·123 of hay ,, 0·008 roots ——— 0·276	400 lb.
At grass grazing, 150 days	9th to end of 13th month	Acres 0·500 of grass	500 lb.
Oats, 600 lb. Oat straw, 800 lb. Hay, 600 lb. Roots, 2500 lb.	14th to end of 20th month	Acres 0·290 of oats ,, 0·200 of hay ,, 0·040 roots ——— 0·530	700 lb.
At grass grazing, 150 days	21st to end of 25th month	0·750 acres	896 lb.

Result at 25 months from 2·40 acres of land, 8 cwt. gross (less calf 90 lb.), or 800 lb. net of beast.

It will be seen from the above statement that this steer leaves a much better return per acre than in the pasture example, i.e. over 300 instead of under 200 lb. per acre. The increased production from the soil is explained in two ways. First, the cow's keep for nearly the whole year is saved; secondly, for over four-sevenths of the animal's life the food came off land that had been worked and had received other expensive treat-

[1] Includes 90 lb. weight at birth.

ment in exchange for an increased output of vegetable produce. There is probably a correction to be made when the yield of carcase percentage is considered and the result taken as meat produced per acre. Most unfortunately there is no direct evidence; for it would appear that no attempt has been made to obtain authentic data on this point. But calves brought up on the pail always seem to suffer from what their owners describe as "losing their calf flesh." An inquiry which was carried out lately by Dr Marshall and myself for the Cambridge School of Agriculture on behalf of the Board of Agriculture clearly shows the chaotic state of knowledge about such matters; but nevertheless it may, I think, be assumed with confidence that the "hand-reared" stores would not yield as well as those brought up in nature's way. If we surmise that the animal starting as a "pail-fed" only yields 51 per cent. and compare that with my assumption (in itself a deduction made from data which are far too scanty to be reliable for anything more than an estimate) we may make the following comparison: the store from the continuous life on the hillside, aftermath pasturage and other grass-land, has been estimated at 54 per cent. carcase to yield 120 lb. of meat and meat equivalent, such as hide, etc., to the statutory acre. The beast whose career has last been tabulated at 51 per cent. of carcase would give a total of 410 lb. meat, plus 100 lb. equivalent, or 510 lb. from 2·4 acres, that is to say just over 200 lb. per acre.

Much the worst side of this animal husbandry, as practised before the war, was the underpaid labour, or rather drudgery, which the care of the young stock in the yards in winter involved. A cowman, rearing the calves, got more, but the feeder of the "buds"[1] and yearlings could not profitably be paid more than sixteen shillings a week, for which he was expected to feed on the Sunday morning and evening as well as work long hours throughout the week. But even on such wages it is difficult to show a profit for anybody concerned—always excepting the drover, the dealer, and the salesman. A 8 cwt. store of this class

---

[1] Calves from seven to 14 months are often so described. In cattle work as in all departments of husbandry the teacher or writer is much handicapped by the absence of any exact nomenclature.

was well sold at £16; I have bought them for as little as £13 per head.

Poor as these results seem, I have "mouthed" very many "stores" on the big markets of England, as well as in Ireland, that weighed less than 6 cwt. at 30 months old. On the other hand, better bred animals that have been done well from early life either on the pail (though this is very exceptional), or after leaving their mothers, will do 8 cwt. at 20 months; but these are less common than the starvelings who go under 5 cwt. when their first two broad teeth show them to be over 22 months old.

From the point of view of feeding ourselves in this country, a great deal too much land is doing nothing but sustain the life of young oxen. Some land, of course, being inaccessible or too poor to work, cannot be used to better advantage. But, whilst fully realizing this, one can but deplore that the economic conditions of the past 40 years have forced many of our farmers to an adventure as unscientific as it is unproductive. For every day of an animal's life which sees no reasonable growth, is waste of feeding material. The animal's body-heat and the working of its internal organs absorb food to no purpose when there is no increase in weight, and to little purpose if the increase is slight. It will be shown later how this is done and further demonstrated how the acreage employed in our two examples can be made not only to grow the same amount of meat, but, at the same time, a large amount of bread.

CHAPTER III

GRASS BEEF

IN the previous chapter we dealt with land which, without deep cultivation or other expensive operations, will produce a very large output of plant life. It is on fields of this kind that our famous grass beef is made prime. No other densely populated European country dreams of grazing so large a proportion of its surface in this way; it is doubtful, indeed, whether any have the live-stock to produce the quality of meat which long experience has taught us to expect, and without conceit we may congratulate ourselves that the Britisher and the Irishman have special gifts for carrying on this class of work. For in very few instances does the amateur deceive himself more than when he assumes that anyone can manage a high-class grazing farm. But we will first consider the soil, the fundamental part of the whole subject.

The soil of almost all the "finishing" land tracts has two qualities that are, unfortunately, rarely given to us by nature in combination; it is fairly open in texture and at the same time composed of material containing large store-houses of plant-food. But unhappily many soils that are of good texture and composition are very far from being good enough simply because of their water content. Graziers—as these summer beef-farmers are called—may often be heard, when describing the beauties of their land, to say that such and such a field is "cool-bottomed." There is nothing mysterious about this description; it simply means that the water-table is near enough to the surface for the soil-particles to lift a film in times of drought, and yet sufficiently deep down to allow of the well-textured soil freeing itself of excessive moisture in times of heavy rain—in this latter capacity it can and, under some circumstances must, be helped artificially by land-drainage. These qualities are very seldom found in combination, and such land, being very scarce, naturally commands a high rent. Compared with store-land it would fetch 100 per cent. more rent; compared with average good plough-land, 200 per cent. more rent—provided it were well watered, moderately shaded and sheltered, and properly fenced. But even with all these, it can easily be spoilt by bad management.

The surface of these pastures has to be supervised and worked so as to keep them well covered with a carpet of grass that is sweet, nutritious and continually growing from April to the beginning of winter. Otherwise weeds will take the place of useful grasses and clovers. Much skill is required of the grazier in this respect; to keep down weeds requires constant care on all land, even on that under permanent grass. The roller and the harrow must be used at the right time of year so as to improve conditions for the more delicate of the valuable plants; and when used with discretion the teeth of the harrow will check, and even tear out, some of the obnoxious ones. Later in the year the scythe will need to be used to cut back such weeds as the nettle, which on good land grows with an astonishing persistence. But, above all, judicious grazing must keep a thick, level, and evenly growing carpet of herbage on the land. It will only be thick when the small varieties—bottom-grasses, as they are called—cover all the spaces on the surface which the tall-growing top grasses leave vacant. To keep the sward level the fields must be so fed that no one particular kind of grass gets too high; on the other hand, it must not be eaten down so close as to damage or uproot any of the small but useful plants. A continuous yield can only be secured when the general good management insures that different varieties of plants with differing periods of growth are present, in their correct proportions, among the flora composing the growing pasturage. When all the management is good there will be a regular rotation of growth; the early growing grasses will be bitten off and succeeded by those that bloom later; while these are being eaten the later growing grasses will be mixing with the second growth of the earlier kinds, and so on. This succession can be helped by skilful manuring; a few of our practitioners know this and take full advantage of their knowledge, but too many are content, in their ignorance of the soil or of manures, or of practical botany, or of all three subjects, to go on without the great help which scientific knowledge can give the farmer.

The grazier must have some very considerable knowledge of stock, as well as of land. He must be good judge enough to buy those that suit his purpose; he must have observation enough to tell almost at a glance how well or how ill his animals are doing while feeding in his fields; he must know, almost as by instinct, how and when to move them from field to field, for the animal's sake as well as for the good of the turf. The change of grazing ground, by moving animals from field to field, is quite important in the process of regular and rapid beef-making. He must recognize by the attitude of an animal when it rises from its resting place, from the carriage of its head or from the appearance of its hide that it is ailing and be able to give it the immediate attention that will probably ward off a serious illness. Finally, he must know, though on this point many fail in good judgment, when the time has come at which it is no longer profitable to go on feeding an animal, the time for one to be sent to the shambles, and another put in the field to clear up all that the fat animal has left uneaten.

One of the great difficulties of such a farmer's career is in obtaining a proper supply of raw material, or store cattle, to grow till they make prime beef. The success, indeed, of the enterprise is frequently endangered not only by the store-stock being priced too high, when the selling-out price of prime bullocks is considered, but also by the inadequate number of animals of good quality. The remedy for this will be discussed later, but we may here refer to one improvement that is quite practicable under conditions similar to those existing before the outbreak of war, should this country unhappily return to them. If improvement is to be made in the matter of increased production, the change in practice here advocated will be necessary for the sake of further guarding ourselves against danger due to dependence upon oversea-supplies.

Graziers like old cattle best; they will not willingly buy animals younger than 30 months, and they are better pleased if the store shows the six broad teeth indicating that he is at least 3 years old. The "full-mouthed," that is, animals of 3½ years or more, are valued most of all. Having given 10 years of close observation to this subject, I feel some confidence in expressing judgment upon it.

Eleven years ago it became my business to study the practice of the graziers in one of the best Midland counties, a famous hunting shire specially endeared to the fox-hunter by its large, level fields of richly-carpeted turf, a country with fine open gallops and wonderful hedges to make jumping difficult. Only soil naturally rich or fertile will grow hedges thickly enough to test the prowess of the really good man, horse, and hound. I was then told, as often before and since, that young oxen could not be kept on this good land, that any beasts with less than four broad teeth "went off" immediately they were turned on to it, that younger animals even died, and so on. Now I might have accepted this without evidence, as I have often accepted the statements of other "practical" men, had it not been for one strange anomaly with which I was familiar on this very type of land—namely the practice of farmers who bred high-class pedigree beef-cattle. The breeders of these aristocratic cattle turned out their yearling heifers, and even young bulls, on to the very best pasture without any ill effect. In fact, while I have never heard a pedigree-breeder complain of his land being too good or his grass too strong for his grazing animals, I have often heard complaints of the reverse. So, after careful observation and experiment, I have formed the following opinion:

It is certainly safer to turn older cattle on to the best pasture, for this may be done without elaborate precautions as to their over-eating, or their being affected by the extremes of weather in the early part of the year. With yearlings (and still more so with "buds") great care is necessary to prevent their "blowing"[1] themselves with the young grass, scouring themselves by eating too much, and so on. In other words, the older animals are easier to manage and cause comparatively little anxiety; the

---

[1] This state, also commonly called "hoven," is known to the veterinarian as *tympanitis*. It consists of the distension of the huge stomach, found in the ox, with the gases generated by the fermenting young grass; in young animals this disease frequently causes death if not attended to quite early in the attack.

youngsters give their grazier much trouble and a considerable amount of anxiety. But if a large portion of store-land is to be farmed, and not robbed, in the future, this care is essential and the anxieties must be faced and overcome. By using younger animals, more will be available from a reduced quantity of land. Under our present system ten acres of store-land will often carry two cows, one two-year-old, two yearlings and two calves. This stocking enables one yearling and one two-year-old to be sent out every year. Put a foster calf on to one of the two cows to be reared with her own, and turn out three yearlings every year, and the store population goes up at once. Rear two foster calves, one to each cow, and turn out four yearlings every spring and the store cattle population goes up still further.

Before going into the detail of this economy, I may say that by personal experience I am familiar with the difficulties and dangers of turning out a bunch of young yearlings about 15 months old on to rich pasture producing quick-growing, succulent grass, especially in mid-spring when the nights are still chilly and the ground is often covered with white frost at dawn. If neglected they will "blow," they suffer from diarrhoea, and above all, they are attacked by the small "lung worm[1]," which causes "hoose"; and young cattle suffer very much more from all these evils than older beasts. But experience has also taught me that the difficulties can, with much care and trouble, be overcome. The farmer grazing this very young kine must first of all grow some hay; he must, further, supply shelters in each field. These need not be elaborate, but they must be substantial enough to keep off the worst of the weather, and to give a dry layer. Each shelter must be surrounded by a small enclosure, or pound, in which the young animals must be shut in at nights and receive a breakfast of hay every morning. If the grass is causing them to scour, they must be given more hay; if the frost has been very heavy they may have to be kept in later in the day. They will pay for some astringent food, such

[1] The *Strongylus micrurus* and *S. filaria* inhabit the air passages, these and possibly other minute worms cause the terrible cough and other distressing symptoms called "hoose" or "husk." Young animals constantly lose health and even die through the mischief caused by the worms. Older bullocks, though often much upset by them, seldom suffer very seriously.

as cotton cake, though this is not necessary if the long fodder is
of good quality. Old cattle can do without all this care: but
young animals will not thrive, some of them are likely to die,
if they do not receive this attention during the first five weeks,
or so, of the grazing season. Undoubtedly they are much more
trouble to their owners and, obviously, they require some extra
expenditure on labour to win the hay and to supervise them with
skill and care.

There is a further point that must be carefully considered.
Starvelings are no good for this work. The wretched little
beast that has been starved on an ill-filled pail and a badly
supplied manger in babyhood, has galloped about hot, burnt-up
pasture from the age of four to nine months, has then gone into
winter quarters in a wet yard with dry straw, a few turnips and
very little hay for food, and so is little more than 400 lb. weight
at 15 months old, is no use to anyone for feeding purposes. If
the grazier is to change his practice the rearer of these starvelings
must be completely reformed. This is one of the difficulties in
all agricultural practice: husbandry is so much of a jig-saw
puzzle that each member's work must be dovetailed together
if a success is to be achieved. The amount of money wasted by
the individual, to say nothing of the loss to the State and the
neglect of production caused by the vast droves of these miserable little bullocks, still often described as calves though about
15 months old, is at present a handicap to every progressive
feeder.

It may be explained how and why the extra expenditure on
labour may be not only recovered, but, as I hold, made a means
of adding further profit to the occupier of the grazing holding
—quite apart from the great object we have in view of getting
more human food from each acre of our Kingdom.

At the present time the conditions of all our farms of good
grazing land are somewhat as follows: it is seldom that the
land of any one farm is all finishing bullock-land, but, for simplicity's sake we will assume that this is so. We will suppose that
the year begins with the coming of the grass, and similarly that
the farmer buys in his cattle gradually so as to be fully stocked by
the time the full flush of grass is on the land, say by the middle

of May, the date varying with the geographical district, the aspect of the land, the nature of the soil, season, and so on.

Each acre of land is expected to carry one large bullock, the varieties chosen varying with proximity to certain markets, the farmer's idiosyncrasy, and various other causes. The relative numbers are, roughly, as follows: Shorthorns (more or less thick-fleshed according to their breeding) are by far the most numerous; next, Herefords, the famous "white-faces" beloved of all grass-land men; after these, though a long way behind, come the quaint black-polled Scots or their crosses, the famous Blue Greys, the strong-boned Black Welsh and Lincoln Red, and, in far smaller proportion, Devons,—though in some districts they predominate—South Devons, much fancied on account of their hardiness, though they are large feeders even for their great size; a very occasional group of the Sussex breed (most level of all grazing cattle, for they eat everything as it comes); West Highlanders, and crosses of all the breeds mentioned.

While the casual observer might say that Shorthorns outnumber all the other breeds put together by two to one, many so-called Shorthorns are of mixed, or even of mongrel, origin. Whatever the breed, they are all large stores, weighing alive and unfasted about 1000 lb., i.e. 72 stone imperial or 9 cwt. on the average. As far as possible, they are all in good health and just in "fresh" condition, their frames, or skeletons, being well covered with flesh, their muscle mixed with little fat, and their digestive organs little, if at all, larded with suet and "fat[1]." It must be candidly admitted that the quality (i.e. their capacity to thrive or put on weight) is a very varying factor. Probably few are really good, and many, it is to be feared, are quite poor specimens of their race. It is indeed a strange thing that this country, the stud-farm of the world, should produce so large a proportion of a grade low enough to assort with the most primitive of their race. The large proportion are steers, though

---

[1] Kidney suet goes with the carcase, "apron" or "caul" or "caul suet" is the butcher's chief perquisite, the "gut-fat" is often of value as suet, but sometimes goes with the "waste fat" to the soap-boiler and some other fat is unfortunately absolutely waste material.

heifers are rather preferred, especially where the grass-land is not quite the best of its kind. Some graziers are very partial to cow-stock, that is to say, they buy up dry cows to fatten instead of steers.

The beasts, of whatever kind, are distributed over the fields and are expected to feed for a period lasting about 20 weeks, during which time each head, on the average, consumes the bulk of the grass off an acre of land and, in so doing, increases a little in size of skeleton and in weight of lean meat or muscle; this we must simply assume, for unfortunately little is known on these two points. They also cover their bodies under the hide with a thick layer of adipose tissue, they fill in the interstices between the muscles with fat, the muscle tissue also becomes infiltrated with fat, that is to say, the meat becomes "marbled," as it is called, and while all this is going on, the abdominal cavity becomes very heavily larded. At the end of these processes the store-bullock becomes a prime beef beast[1].

The gross increase in weight made by these grazing beasts has been assumed to be 20 imperial stone live weight: on the average there is, indeed, some evidence in support of this assumption in the weighings taken by Mr C. B. Fisher, of Market Harborough, and published in the Royal Agricultural Society's Journal for 1894[2]. It appears to me, however, from many observations and deductions, that, if no cake is fed on the grass, this assumption overstates the increase made by the animals; and that if a sufficient number of weighings of good, bad, and indifferent animals were taken over several seasons,

[1] The matter of the usefulness of this increase of fat for human consumption is an important one. Dr F. H. A. Marshall and I carried out an investigation on the subject for the Board of Agriculture and Fisheries during 1917. A brief résumé of this work was published in the Board's Journal for September, 1918, but the Government has not thought well to publish the whole report. The numbers of animals we worked upon may not have been great enough to establish the most reliable data upon which to build knowledge, but our figures, if scanty, most certainly established the need of reconsideration being given to the whole matter. We believe that our figures show that the whole of popular belief about the matter of fattening cattle is founded on a misconception as to what really takes place.

[2] *Anomalies of the Grazing Season of* 1894, p. 667 et seq.

the results would show an average increase more like 16 than 20 stone. However, before the war, it was quite unusual for a farmer to weigh his cattle either at the beginning or at the end of the season, so it did not matter to him how much weight they put on. The only way in which 99 per cent. of the owners could calculate was by comparing prices of store-stock when bought in and the prices made by the finished stock at the end of the grazing season.

Before 1900 I was frequently told by friends who were experienced in the work or by those who were familiar with the trade, that an increase of £5 per head was necessary for the grazier to get a good living; and between the years 1890 and 1900 I believe that most men either did this, or failed altogether. From that time onwards the cost of stores went up without any corresponding rise in the prices of beef, and the rents, the big item of expenditure after the store had been paid for, were about as low as they could be squeezed. Thus, in 1908 and 1909, as I learnt on good evidence, it was only the exceptionally good man who got an increase in value of £4 a head on the average of the stores he fed, and of this sum £2. 5s. was rent. Over and above the grazing season's grass, the land provided something for the keep for the rest of the year; some farmers would run a ewe-flock very thinly spread over the land, others a few colts, others a few store-cattle, others all three. But, even with this help, the farmer's living could not be said to be a fat one[1].

It is difficult to calculate with uncertain data, but the following account of the grazing of big three-year-olds and the finishing of yearlings may be taken as accurate.

Five yearlings, weighing about 600 lb. each, could be kept on the same area of grass as would keep three big bullocks of 9 cwt. in weight. By all the available evidence (which is to me conclusive) each of the five young animals would make at least the same gain in live weight as each of the three older bullocks —provided that the precautions previously referred to are taken. The cost of producing the young stores, though they involve

---

[1] It is hard to foresee this season (1918), when stores cost well over 80s. and feeders will have to sell at about 75s. per cwt. live weight, how any better financial return will be obtained.

more trouble, is less in feeding stuff per unit of live weight than is the case with the six-toothers; further, as it has been shown that more of them would be available, it should be possible for the feeder to buy at a slightly lower rate and yet leave at least the same profit for the rearer. As regards realizing the profit, the relative price paid per cwt. for large and small varies considerably. When large prime bullocks are scarce, they make as much per stone, sometimes even more, than small young prime animals; but usually the reverse is the case.

From the national point of view the estimate of the yield of carcase is a fairly safe one, and other useful stuff may be reckoned in the same proportion, except that the hide of the large beast will no doubt be considerably more valuable. The yearlings, decently brought up as suggested in a former chapter, may be estimated to yield from 52 to 55 (say 53 per cent. on the average), and the three-year-olds about 3 per cent. more. If, as one hopes, the whole question is in the future submitted to the proof of systematic research, this difference will probably be found to be exaggerated. But, given this difference, the result will be as follows:

If the three-year-olds yielding 53 per cent. (this low figure is deliberate) raise themselves from 9 cwt. to a weight of 11·5 cwt. at 56 per cent.—a total carcase receipt of 721 lb. (less 534 lb., the carcase weight of the store)—we get a total yield of 561 lb. as increase from three bullocks off three acres of land. With five yearlings, assumed to raise themselves from 5·5 cwt. at 50 per cent. carcase to 8 cwt. at 53 per cent., we find in the same way a difference of 167 lb. per beast; but five would be grazed instead of three, and the amount off the same area of land would be 835 lb. There is a further correction to be made. Dr Marshall, working on the inquiry already referred to[1], obtained results which show that these small animals will probably yield a larger proportion of gristle and bone. Assuming his figures to hold good, the difference is, however, under 3 per cent. and the increase shown above would be only reduced by 25 lb. on the whole three acres —leaving a surplus of 230 lb. on the five young animals.

The system advocated here has the further advantage of

[1] See footnote 1, p. 30.

giving more employment on these tracts of rich grass-land. The supervision of the young cattle in spring and the hay-making in early summer, would ensure the employment of extra permanent hands, for, as it is now, the number of agricultural labourers employed is often negligible. I have known the staff on 350 such acres to consist of one old man, while even the farmer himself, though constantly working on the farm, lived in a neighbouring town. A further advantage of having more hands would be better surface-tillage, more weeds cut, ditching, hedging, and draining improved; for, during the last 25 years, all these have gone back on many holdings and so the land has produced less. An increase in rural population must always be the aim of anyone who wishes to benefit his country through the land. If the adjacent lands of inferior quality—and there are nearly always some such—were ploughed up, instead of being, as they often are, left with a miserable covering of grass or weeds, the conditions would be easier.

But to make things as perfect as may be, this splendid foundation of an almost ideal agriculture should be combined with some rural industry that would find winter employment. England is a beautiful country and no one could wish that its fine grazing areas should disappear; the authorities therefore should make every effort to preserve their good qualities and to secure improvements in them. Who are more suited to this task than the owners themselves of the broad acres, noble trees and rich pastures which are the finest of their kind in the world?

CHAPTER IV

WINTER BEEF

IT may be said without hesitation that no practice is more typical of empirical British industry than that of winter beef-production. Its produce, "Roast Beef," is supremely typical of our home life, and yet, wheat-growing possibly excepted, few items in our agricultural practice have brought more men to financial ruin. I have, as it happens, known many who, being in a very humble way of life, have told me of the glories of a father or uncle who farmed in the good old style. Nearly all these poor friends of mine—I use the adjective in its literal sense—have told me, with obvious delight, when speaking of the past, of the fat bullocks that were part of their family's pride: very little inquiry led one to see how great a share these delusive, if attractive, creatures had taken in the change in the family fortunes.

In the whole vicious cycle which unsound economic conditions have forced upon our farmers, there is nothing more disastrous than the methods prevailing among many who pride themselves on the quantity and excellence of their winter beef. Among these there are many who pay no attention to the help which science has to offer; and their conceit in their own success prejudices them against a test, by proper accountancy, of the money wasted in this, and other branches of their industry. But despite their belief in their efficiency as practical men, they fail, most of all, in their unbusiness-like way of marketing their goods. Close observation of the markets for about 20 years brings the conviction that this class of beef might have been made profitable to grow, had it not been for the over-supply resulting from the carelessness and self-satisfied ignorance of those *who produced it regardless of cost.*

The very best winter beef is skilfully fed with unstinted amounts of "roots," a small amount of good straw and hay, and, in some cases, most extravagant rations of cake and corn; in all cases the concentrated food must be ample. Animals, with a certain age on them, fed in this way for some 20 weeks, produce the very best joints that come to the block between December and May (both months included). No other country in the world produces this meat in any quantity. It is doubtful, indeed, whether any country produces such perfection as regards appearance, delicacy of texture, and flavour. In England such beef, whenever possible, is called "Norfolk," no matter in what county it may have been finished; in other parts of the kingdom it is called "Scotch." The name depends on the side of the border which had the opportunity of housing the bullock, but both titles are a mark of distinction. The superiority of this beef was such that it would always fetch a decent price, even before the war when prices of ordinary food were low—low enough, indeed, to lead to the waste which threatened our security as a nation long after the war began. It was, in fact, a luxury of which the supply might have been marketed at monopoly figures, if the winter feeders had taken pains to regulate the demand, for it had no overseas competitors. But the winter feeders despised co-operation and our co-operative societies were, it is to be feared, too much concerned in saving the commission of half-a-crown on buying a ton of cake, to give any real attention to the marketing of millions of pounds' worth of cattle. The whole industry suffered from lack of business-like co-ordination of interests—from the calf dropped in the west or north country auction shed full of magnificent deep-milking cows to the long lines of splendid, fat, three-year-old bullocks standing in the East Anglian sale-yard. Some form of co-operation is a prominent feature in the agriculture of our Continental neighbours, whose husbandry is a vital part of their national life instead of being generally regarded, as was ours, as an industry unworthy of a good Britisher's enterprise. In the fat cattle trade there was one great exception to this lack of business-like co-operation, not, however, among the farmers, but among the Scots dealers! The business men of the North-East of Scotland had, before the war, a very fine trading net-work for distributing their cattle systematically to those markets where they were most likely to sell profitably.

This must not be regarded as disparagement of the salesman's craft; on the contrary, it cannot be too strongly emphasized that a good middleman well earns his place in the world. But there were too many middlemen earning a living out of the cattle trade before the war; many instances could be quoted of too many small commissions being taken off each beast between the producer and the consumer. Between the departure of the animal from his homestead and the final transaction between the retail butcher and his customer the producer paid money to one man for selling, to another for buying, to another for "leaving" a bullock—but it was partly the producer's own fault.

Having for 20 years before the war had good opportunity to study cattle-markets, a study which gave me much interest and pleasure besides satisfying a legitimate professional curiosity, I have some confidence in criticizing the many practices which, it is hoped, may be improved in the future. Over and over again I have gone to markets to find the primest "Norfolks"[1] supplied in such large numbers as to force their price down to the level almost of inferior, or of "chilled," or of "port-killed" beef. Many cooks and housewives have no real knowledge of the quality of meat and so the number of customers willing to pay for the best is strictly limited; further, the *large* joints from these beasts are not sizeable for small families, so that the real demand came mainly from good restaurants, from county and railway hotels with a reputation for good food, and from the tables of such owners of big houses as had enough money and knowledge to insist upon being served with first-quality joints from the "sides" of large, prime bullocks. Such customers did not of course pay more than they were obliged to pay for the goods they required. When the supply of meat was greater than their customers' demands, the butchers were too business-like to give more than the price current for all classes of meat, good,

[1] I think there may be a double reason for the fine fat beasts being often called Norfolk bullocks. In the first place that county sends them out in great quantity and of superb quality; also, they are particularly the produce from farms on which the "Norfolk" rotation is practised. This system of farming imposes 25 per cent. of "root" crops on the land, and the greater part of these is consumed by the large bullocks in winter.

bad, or indifferent. At certain seasons the numbers of prime bullocks sent to the markets were altogether excessive; consequently the price realized for their carcases was for years not only ruinously low, but was very little higher than the figure quoted for inferior beef taken from cold storage.

On the other hand, one could very occasionally go to a market where the supply of prime animals happened to be small enough to make the connoisseur pay a proper price for what he wanted. Then one realized the folly of flooding the market. No buyers offered one another a trifle on each animal to "leave it alone"; there was no pre-arrangement of the destination of the various lots; no animals were allowed to go through the sale-ring unsold so that the buyers might offer their own fixed price to the unhappy feeder.

I vividly recollect such a scene during a period when the best beef had been selling at a price between 6d. and 7d. per lb. for several weeks. The market-stalls, quite accidentally, contained only *dozens* of prime animals instead of the *scores* which had been on offer for many weeks previously. The price, naturally, went up to over 9d. per lb. and I heard the buyers say, as I have many times heard them before and since, that they must have some first-quality meat, whatever the price. On this particular occasion, as on several others, I carefully noted the prices generally prevailing on other big markets and there was no rise from the general dead level of ruinous figures. How ruinous this level was as regards profit on growing meat the following figures show:

Cost of feeding—Winter beef.

Taking my evidence from some 200 bullocks fed in different parts of the country, I find that it takes on the average 16 weeks' fattening to make a moderately good store into meat, that during a feeding period of this length one may rely on an increase of 2 cwt. live weight, or an average increase of 2 lb. a day.

The figures for some 80 beasts fed till they were "prime Norfolks" show a period of 20 weeks to be necessary, during which time they did not increase quite as much as 2 lb. per head per diem, in spite of the following substantial average daily ration:

Concentrated food (½ linseed, ½ cotton cake)	8 lb.
Cut fodder (⅓ straw, ⅔ hay)	8 lb.
Roots (partly swedes, partly mangold)	112 lb.

A very long experience of watching cattle fed on the above rations leads me to believe that, besides this ration, a varying, but very considerable, amount of the straw supplied as litter is also eaten.

It is a very moot point as to how much of the increase in live weight is carcase. It is an extremely important matter and one which demands immediate and thorough investigation. Lawes estimated that one might expect 80 per cent. of the total increase in live weight to be returned as carcase. I myself venture to predict that this percentage would, on investigation, be found to be too high. Still, assuming that it is so and also that the prime Norfolk gives an increase of 2 lb. a day (though it will be found, in practice, that it fails to do so by a small decimal), we get the following result: *It takes 4 lb. of mixed cakes, 3 lb. of hay, 1 lb. of straw and 56 lb. of roots to make 12·8 oz. of prime beef.* Besides the meat there would be a considerable amount of fat in the offal. This offal fat, though costing the farmer dear in feeding-stuffs, is looked upon by the butcher as one of his perquisites.

The above figures and deductions appeared in my article in the *Journal of the Board of Agriculture* (1908) entitled "The Cost of Producing Winter-beef." Not only have these figures and statements been passed without serious challenge, but they received, by a strange coincidence, confirmation in the results of an investigation[1] carried out in Norfolk itself at the very time when I was lecturing to Farmers' Clubs on the basis of notes afterwards used for the article.

Under pre-war conditions the only means of making a profit lay in the extra richness imparted to the farmyard manure. This idea was developed by all concerned and by no one more than

---

[1] See the article "The Cost of Winter Grazing in East Norfolk" by the Rev. Maurice C. H. Bird in the *Journal of the Royal Agricultural Society of England* (1909), page 82.

the dealer in fat stock; during my lectures in the winters of
1907–08–09 I was constantly heckled on this point. But this
plea is about 50 years behind the times. It held good when
we had to rely solely upon farmyard manure to enrich our lands
under corn, but not since we have begun to understand the
subject of fertility better. We want farmyard manure to improve *texture*, and the humus it contains is often useful in
furthering the growth of beneficial micro-organisms in the soil;
we want it, of course, to help prepare the land for the sowing
of the seed, or, in other words, to improve tilth; the plant-food
it contains is also useful, but it is in no sense indispensable
when a supply of concentrated fertilizers is available. From
the farmer's point of view it is foolish in the extreme to put
plant-food into the land through the cake-bill when it can be
obtained much cheaper direct from the manure-merchant. The
intelligent and industrious farmer has in the past aimed at
getting the largest possible amount of humus and plant-food
into the soil at the least cost. The extravagant feeder of prime
bullocks, on the other hand, simply aimed at getting both these
by means of his beasts regardless of cost. When the corn land
was very good, some personal profit was eventually secured
through heavy yields of grain; when the soil was only of medium
or poor quality, the practice had to be discontinued or the man
working for a living went to the Bankruptcy Court. The only
exceptions were those who could afford to spend money on
farming, among whom there were, most unhappily for the
industry, many landed proprietors. It has been heart-breaking
during the last 20 years to see the bad example set in this
respect on the home-farms of very many to whom one might
have looked for improvement.

In the article quoted above the cost of hay was taken at £3
per ton, yet in my hearing the agent of a prominent landlord in
the Eastern counties boasted before a company of farmers that
although he could sell his hay at £7 a ton, his fat bullocks *were
bound to have* all they could eat, at whatever cost. In the third
year of the war I heard the agent of another landlord in the
West country boast on the market that he had been feeding
500 lb. of best cake a day to fifty 80 stone Devons while grazing

on the best valley grass! Such instances could be multiplied *ad nauseam*; and the great pity of it is that not only did the employers of such agents allow these practices to continue, but actually believed, in their ignorance, that the exhibition of these wastefully fat brutes was doing good to agriculture.

Absolutely the reverse is the truth. Even if there were some special value in the plant-food contained in rich, cake-fed farm-yard manure,—which, with very few exceptions, there is not—it is bound to be a very wasteful way of supplying such material to the land. For the richer the feeding, the greater is the proportion of plant-food that is found in the liquid which often runs to waste. Our farm-buildings are very often structurally deficient of means to collect the liquid manure, and as the convenience of farm operations often makes it necessary to store the stuff in the field, much of the liquid is lost and the essence of what the rich cake-feeding gives to the manure simply goes down the ditch to contaminate the horsepond. I have known one of the most self-satisfied and extravagant winter-feeding farmers cut a channel from the feeding-shed to the nearest ditch, so that the overfed bullocks might "lie more comfortable," thus losing the benefit of the rich fertilizing material. Such deliberate waste is not uncommon, for many of the practical men who feed heavy rations to bullocks refuse to believe in the value of the liquid from the feeding-stall or court.

There is, indeed, great need for a series of demonstrations all over the kingdom to show the value of the liquid part of farmyard manure. It is a matter of common knowledge to the professional agriculturalist, but not to the practitioner. Such demonstrations cannot be given without expense, and in the past the British public, to say the least of it, has not taken a wide view of spending money on agricultural propaganda. The misfortunes of the past four years have, however, been so largely caused by ignorance, that it is not too much to hope that it may before long be thought worth while to spend money in giving information which may prevent waste even of food-producing material. Every penny spent on cake and other feeding-stuff is waste if it is not returned through the animal or the plant; and it may be national waste of the most pernicious kind, since much of the concentrated feeding rations comes to us from tropical countries in whose well-being we have little concern. Thus we pay money to foreign countries for material which through ignorance we partially forfeit without any return —in fact, the lost material causes considerable pollution of our country side.

To instruct a community that is inclined to cling to existing practices and carry conviction in the face of the national characteristic of unbelief in scientific research requires demonstration work carried out at home. It is of little use to explain the methods of Germany or other countries; the public must be prepared to spend money on many demonstrations carried out at the very doors of our farm-houses. On the other hand, if the public pays the piper, it has a right to call the tune, and to insist upon the best use being made of our agricultural land. This cannot be said to be done so long as more feeding-stuff than is necessary is used wholesale for the production of winter beef.

There will always be special types of farming in which the making of very rich farmyard manure may be justified; to these should be left the manufacture of the very fine meat for which there will always be customers ready to pay a profitable figure. If in the future the over-production of the past is avoided, our Christmas beef will become an ornament, instead of an encumbrance, to good English farming.

CHAPTER V

BEEFLINGS

CRITICISM of agricultural methods should be constructive as well as destructive; and I am convinced that the production of "baby-beef" from beeflings might be made to play an important, almost vital, part in the husbandry of the future.

It has everything to recommend it: it has, in the past, been advocated by many of our leading authorities; at the present moment some of our best practitioners produce it; and it enables us to take advantage of all the sound information that scientific research has placed at the farmers' disposal. The objection to it, however, has for the last 40 years been insuperable; for it gave little, if any, opportunity for land-robbery. While it did not pay, on the majority of holdings, to farm intensively, since the extra produce cost more than it would fetch, average land was only profitable when the least possible labour was expended upon its produce. The grass and winter beef, discussed in the last two chapters, concerned animals that had lived a long life before coming to the fattening period, during which little except rent had been spent on their upbringing. It may be said, in fact, that they represented human food produced from the land with the least possible labour. With meat obtained from beeflings the trouble of calf-rearing is no sooner over than the labour of the finishing period begins. This difficulty has been pointed out to me over and over again during the 15 years in which I have advocated the system. Often, when I have urged upon a farmer the advantages of turning out 20 to 30 yearlings every season instead of a dozen three-year-old, he has replied that the calfhood gave more trouble than all the rest of the animal's life; and this is to some extent true.

An animal that is to yield at 10 or 15 months old a carcase of meat fit to cut up into presentable joints must be cared for all the time; you cannot turn such beasts on to the land and leave them to fend for themselves; undoubtedly they "cause trouble" or, in other words, cost an appreciable sum for labour. But the saving in fodder and concentrated feeding-stuff is very considerable and less of the produce of the soil is wasted in keeping the animal alive. This is the crux of the whole matter. Is it to be worth while working the land, or working on the land, to make the produce greater than when it is left alone? If so, produce must realize an economic price.

A mature ox that is not growing heavier is using a lot of food to keep itself alive even if its exercise is confined to getting up and lying down on its litter. It is the same with a young animal, if it is not growing heavier and fatter. It may be even worse than this; for if a young animal is not getting heavier, it may be going back. To deny that this often occurs in stock-keeping as carried on for the last 30 years is to show ignorance of common facts. Fortunately, very many farmers avoid this worst extreme of bad management, and all know that it is bad practice. On the other hand, it is a common thing to find whole tracts of country on which the great majority of cattle are giving a very small return in growth for the very large proportion of food used for maintenance.

The object of baby-beef production is to get all possible growth, at a cost compatible with economy, while the animal is young. Obviously, a frame weighing 400 lb. will not need as much nourishment to maintain warmth and other bodily requirements as will one weighing 800 lb. Furthermore, youth is the period of rapid growth. The older an animal becomes, the less it is able to increase its size. Age deprives it of the power of converting food into growth. The mature animal consumes a great deal to keep itself alive; if there is a surplus, a small portion is stored as fat—nature's provision of fuel against the time when starvation might otherwise deprive the whole animal of its maintenance ration—but the great bulk of any surplus that greed may lead the animal to swallow is passed through the system without doing any good to the consumer. In practice, probably very few of our steer-cattle ever reach the point when growth ceases altogether. We know that complete dentition is not reached till nearly the end of the fourth year. After that age there is little recorded evidence to show what happens, but from personal observation of cows and bulls, it may be said with confidence that some little growth of frame and muscle continues for at least two years after the animal has a "full mouth." On the other hand, there is plenty of evidence recorded to show that the rate of growth decreases very much during the first three years of life; the calf making more growth than the yearling, the yearling than the two-year-old, and so on. These facts are well known to physiologists; the difficulties lie in taking advantage of them in practice.

To obtain the best, or even reasonably good, results requires much care, for the young animal has not the cast-iron digestive system that the old cow or ox, judging by observation, seems to have. The immature beast must be given suitable food or it ceases to thrive, and that is fatal to good beefling meat production. Again, care must be supplemented by skill in the choice of feeding-stuffs, or the expenses will be so high that no financial advantage will accrue to the farmer. This latter point, however, may be said to be secondary, for the very best food is often wasted through want of care. It is at the outset that the problem of beefling production is most difficult, for after the first three or four months, or after the weaning has taken place, the difficulties are very much less, though care and skill are, of course, required all through the animal's life.

The first consideration is that of milk. A calf running with its mother will consume from 150 to 400 gallons; these figures are only estimates, for little attempt has been made to record how much milk is given by cows of the beef-breeds—and it is with the beef-breeds that this class of rearing most often takes place. It is, of course, admitted that sucking the mother is by far the healthiest method of feeding for the young calf, but it is apt to be altogether too extravagant a system. It has been shown that to keep a cow a whole year for the sake of one weaned steer calf may be a form of land-robbery that only the most unhappy state of agriculture can justify as a general policy. But the fact remains that a calf brought up on its mother, under good conditions, till the age of four or five months has an ideal start in any career concerned with beef production; heifers

wanted for milk production are liable, by this method, to have
their milk-secreting glands overloaded with fat, to the detriment
of their profitableness in later life, but for growing meat it is
ideal. It has already been shown that, in our variable country,
locality may, to a limited extent, justify keeping one or more
cows to yield one weaned calf as the result of her year's keep;
and it might be contended that farmers, whose land is situated
under such conditions, might do well to turn over the produce as
baby-beef rather than as store-stock. But we are now concerned
with producing meat from plough-land, and this entails a far
greater output of beef from each acre. In fact, the cow kept for
the breeding, and rearing, of one calf represents the very lowest
type of production from the land. It is my object to show that
beefling production can be carried on with the most intensive
production from each acre of land held. The Danish farmer
combines intensive arable land-farming with milk, veal, and
pig-meat production; in our future agricultural campaign we
should do well to add baby-beef to the list. To do this to
financial advantage we must decide what is the most economical
amount of milk to allow the calf.

In considering the problem last stated there are two main
systems that merit careful examination : (1) the rearing of several
calves on one cow, (2) the rearing of all the calves, say from
birth or at any rate from the fourth day, on the pail or, as it is
called, by hand.

The first has advantages. I have myself practised it very
successfully with good deep-milking cattle. With three cows
yielding an average of 800 gallons of milk, we reared 24 calves;
three went for veal or were kept for bulls, 14 were brought
out as baby-beef, the rest—heifers—were kept as breeding
cattle. The management necessary to do this demands a little
intelligent care in getting the cows to take to the foster-
calves, and judgment in letting the different calves suck the
proper and economical amounts of milk. With such deep-
milking cows, directly the calving troubles are over, a second
calf should be introduced; then, after about three weeks, a
third. An appropriate method of management, which is known
from experience to be sound, is set out in the hypothetical

calendar here reproduced as a guide to those who have never practised the system:

Weeks after calving	Own calf	2nd calf	3rd calf	4th calf	5th calf	6th calf	7th calf	8th calf
1st	heifer	2nd calf						
2nd	do.	steer	3rd calf					
3rd	do.	do.	bull					
4th	do.	do.	do.					
5th	do.	do.	do.					
6th	do.	do.	do.					
7th	do.	do.	do.					
8th	do.	do.	do.					
9th	do.	do.	do.					
10th	do.	do.	do.					
11th	do.	do.	do.					
12th	do.	do.	do.					
13th	Weaned	do.	do.	4th calf				
14th	suckling	do.	do.	steer	5th calf			
15th	13 weeks	Weaned	do.	do.	heifer			
16th		suckling	do.	do.	do.			
17th		13 weeks	do.	do.	do.			
18th			do.	do.	do.			
19th			do.	do.	do.			
20th			do.	do.	do.	6th calf		
21st			Weaned	do.	do.	steer		
22nd			suckling	do.	do.	do.		
23rd			19 weeks	do.	do.	do.	7th calf	
24th				do.	Weaned	do.	steer	
25th				Weaned	suckling	do.	do.	
26th				suckling	10 weeks	do.	do.	
27th				13 weeks		do.	do.	
28th						do.	do.	
29th						do.	do.	
30th						do.	do.	
31st						do.	do.	
32nd						Weaned	do.	8th calf
33rd						suckling	do.	steer
34th						12 weeks	do.	do.
35th							Weaned	do.
36th							suckling	do.
37th							12 weeks	do.
38th								do.
39th								do.
40th								do.
41st								do.
42nd								do.
43rd								do.
								do.
								Weaned
								suckling
								12 weeks

The outstanding advantage of this system is that the calves consume the milk in what is by far the most wholesome way; they therefore derive more benefit from the food, run less risk of their digestions being upset and the danger of their becoming infected by disease is reduced to a minimum. The domesticated calf is very liable to infection, of very serious import to digestion, when gulping down milk from a pail, and the baneful microbes seem to be very much better kept in check when the process of suckling allows only small mouthfuls to be swallowed at a time. On the other hand, the number of calves reared from the cow is not altogether satisfactory. Even supposing that such a cow was grazed through the summer five months and that she reared five out of eight calves, while at grass, as has often been my experience, an animal giving 800 gallons requires a large quantity of nourishment. Supposing the cow to be housed the whole year the produce off 2·5 acres of plough-land at least would be used in growing the eight calves from 4 to about 100 days old. It may safely be estimated that for the 13 weeks' milk the calf would give an increase of 125 lb. When thriving, the little beast would increase more than this—he ought to do about 200 lb. in the time—but I have reckoned that the other foods he ought to have would be used to create the 75 lb. that differentiates the two weights. Eight calves increasing in weight 125 lb. on 2·5 acres give a production of 400 lb. per acre; a return very different from the paltry one obtained from land that is merely being used as a breeding-run as shown on page 18.

This method requires ample labour and almost the same accommodation as where the cows are milked by hand. It is well if each cow has a box to enter at "milking-time" in which the calves live while their parent is out in summer or while she is "yarded" in winter. The calves can be kept tied to the wall all the time, and **should** be tied up for some little time after feeding on the cow; otherwise they soon learn to suck each other. The man in charge must use some skill in making the cows take to the foster-calves, in seeing that each calf gets enough, but not more than enough, milk, and in supplementing the milk by suitable food and water when the calf has passed the age of about four weeks. The growth of calves, in my experience, is often retarded through their having too little to drink. When their milk has been curtailed, the little animals do not eat enough dry food to promote proper growth and their fluid allowance must be made up to two or three gallons a day according to their size; this at first should be given at a proper temperature and as regularly and punctually as all other nourishment. The foods are similar to those to be mentioned later. Now and then a cow shows herself very troublesome in taking to foster-calves. The attendant must then stand by and restrain her, as far as possible, from kicking; but when the cow is tied by the head it is wonderful how skilful a calf becomes, once it is a few days old, in avoiding her unkind attention while feeding, and it is still more wonderful how little harm is done by any kicks that may unfortunately get home. All this is light work, requiring skill and attention rather than hard labour, and the feeding, management, and watering form suitable employment for women; the cleaning of the boxes (and calves should not be kept with the same accumulation of litter under them as is allowable with older animals) is harder and demands the services of a man. It is, of course, quite possible to have the cows tied up in stalls in a cow-house and lead the calves to them, but this, though it may save some little structural accommodation, adds considerably to the labour bill. I have, however, seen the system successfully carried on under almost all conditions of housing.

The other method—pail-feeding—can be successfully practised with a much smaller allowance of milk. With skilful calf-rearing I know, by experience, that even 10 gallons of milk, over and above the colostrum given by the mother during the first three days after calving, will just give a start to a successful beefling's career. But though this may be done by one specially endowed with the qualities necessary for calf-rearing, such people are exceptional. On the other hand, an allowance of 50 gallons of milk is sufficient to enable the average man to rear calves which will make good carcases of baby-beef at the age of 12 or 15 months. It is a fact, though the inexperienced may doubt it, that very good calf-rearing is a special gift, akin to the genius of the artist! Though the food used may be the same both in quality and in quantity, some will produce plump, sleek and thriving animals while others will turn out a bag of bones covered with a scurvy skin. But all, by a little attention to intelligent instruction, may learn a few simple rules that will help them to rear good calves on moderate rations; and 50 gallons of milk is ample, if the following rules are observed:

(1) The milk should be fed before it gets too cold. If it has to be warmed up or if it has to be diluted (and it is a good plan to dilute it gradually till the calf gets accustomed, by degrees, to drink chilled water in place of milk), *on no account should it be fed too warm*. The milk drawn from the udder of the cow by the calf is at a temperature of 100° F. It is wise to feed the milk from the pail before it has fallen below 90° F., though no harm will be done if it falls a little below this point. On the other hand, a rise to an appreciable degree above 100 is very harmful indeed. I have investigated this many times by following good and bad rearers about and unexpectedly testing the porridge, or milk and water, or other fluid with a thermometer. One of the most valuable gifts of the natural calf-rearer is a sensitiveness of touch as regards temperature. But with the help of a thermometer anyone of average intelligence will soon learn to feel when the food is too warm.

(2) Cleanliness is at least as important as right temperature With a little care and a minimum knowledge of the fluid dealt with, there should be no trouble about this when milk and water only are given. For reasons of cleanliness alone, I am opposed to any form of gruel for calves being reared by hand for purposes of meat production. Such stuff as the average "calf meal," or milk substitute, sticks to the pails, and experience shows that so much labour is involved in washing and making them sanitary that they are "not worth the candle." In districts where milk, at the rate of from 40 to 50 gallons per head, is too expensive to be given to calves wanted eventually for beef, rearing will not pay the owner unless he has a particularly good feeder in his service. But this will be dealt with in another chapter. As regards the use of milk, there are two common causes of failure—neglect, or ignorance, or both. If the pails are not rinsed with clean, *warm* water *directly* they have been used, the dregs of milk are contaminated by disease-producing germs which, in the warm atmosphere of the calf-house, multiply greatly and are ready to invade the whole of the next

meal of warm milk. Failure to rinse out the pails at all is as frequent a source of mischief as it is inexcusable, but sheer ignorance is often another cause of trouble. Boiling water coagulates the milk; and, if the pails are "scalded" before being rinsed with cool water, the cracks and joints in the vessel get full of a jelly-like substance which in time becomes a hot-bed of living organisms, often of a malignant kind. A thorough cleansing of the pails once a day with *boiling* water is desirable, but, before it is done, the utensils should always be rinsed with warm or cold water. It is astonishing how ignorance in this respect does mischief on the farm.

(3) Every one of the farm hands employed in the industry would be more valuable to his employer, and more interested— and therefore happier—in his work, if he were taught the first principles on which his work is based. But the rural school technical educationalist would seem to have confined his instruction, in the past, to gardening. Neither in their school years nor afterwards has there been a systematic effort to awaken an intelligent interest in children's minds in the structure and function of the animal body; to give them a reasonable grasp of the composition of foodstuffs; or to train those who have to work among them in the observation of the wonders of nature which enable the earth to produce human food. Yet, without this sort of knowledge one cannot hope to obtain a prosperous agriculture. Without intelligent appreciation of the mysteries going on around him, the worker on the farm is as great a drudge as any in the city, and work in the country, without a realization of its beauties, will always drive good men to seek the consolations provided for those who labour amidst brick, stone and iron in the polluted atmosphere of our manufacturing districts. An intelligent appreciation of what one is doing and *of why it is done* must be brought into the art of calf-rearing if success is to be achieved, and, unless some such education as is suggested above is provided, there is little hope of improvement in cattle husbandry—or in any other agricultural pursuit.

Having fed the 50 gallons of milk, together with other suitable foods, we must consider what return may be expected from the milk of an 800-gallon cow by this form of meat-production. From the exact figures of calves so reared I see that each calf made a gain, by the help of the milk, of 70 lb. during the period of ten weeks' pail-feeding; also that 65 lb. of the gain may be credited to the milk, so that 16 calves would give a total of 1040 lb. Allowing, as in the previous example, that the cow supplying the milk uses the produce of 2½ acres we get rather more weight of calf—416 lb. as against 400 lb. per acre of plough-land.

This extra amount of produce is not, however, the main advantage of this last system of calf-rearing, for it must be remembered that on plough-land one of the objects in mind is the making of farmyard manure, and from this point of view it is desirable to raise the greatest possible number of good beeflings per head of milch-cows available. The pail-fed calf also needs less housing accommodation. Ten cows standing in a cow-house to be milked, with the calves they rear kept in boxes by themselves, take up less room than that required to house each cow in a separate box, and so the capital needed to work the farm is reduced; though here again this is not a very important consideration.

The choice between the two systems—and there is no reason why they should not be carried on together—will probably be decided by convenience or even by personal preference. In some cases, the necessity of milking the mothers by hand is a serious disadvantage of the second system; in others, buildings will decide the choice, and so on. But that the return for food consumed is greater with beeflings than with older cattle is evident from the following figures taken from actual practice. The animals described were the produce of cows, mated with good Shorthorn bulls and kept in ordinary dairy herds where milk-selling was the only object in view; the calves were removed from the milk-producing farms to ordinary arable land when only a few days old. The statement is published in the certainty that under ordinary good management the results can always be repeated with decently bred calves.

Allowing a beefling 50 gallons of milk, to be hand fed, the animal can be marketed at about 52 weeks old weighing about 6 cwt. live weight. The percentage of carcase weight yielded will vary considerably, probably according to the quality of the calf, but may be taken at 54 per cent. To produce each hundred-weight of animal the following foods will be used:

Concentrated Feeding Stuff	182 lb.
Hay	336 lb.
Good straw (partly as chaff)	112 lb.
Roots or Green Fodder crop	896 lb.

While the animal is still young, linseed cake, bran and oats will be the staple concentrated food wanted. After the first four months these may be gradually replaced by less refined and more concentrated foods. Skill in making up the ration must be assumed or the above results will not be obtained. If the calf is well littered it is likely to eat a lot of its bedding, when it will be found that the above-mentioned straw may be wholly or in part omitted from the food given.

One particular objection has always been raised against baby-beef, namely the difficulty of providing fodder crops for the young animals in the late spring and early summer months and the cost of green-soiling, or cutting and carrying home the green crops when grown. Now, these difficulties are exaggerated. The management of any enterprise must be such as to allow labour to be available for any profitable undertaking, and skilful foresight can always supply the succulent vegetable food that is a most desirable item in the dietary of young cattle being forced out at an early age. But, as I know by experience, it is perfectly easy to do with grazing instead of green-soiling, if desired. Well-shaded paddocks are all that is necessary. I myself have always used grass-land under a permanent sward, but there is no reason why young cattle should not graze any field under a suitable seeding of rotation grasses. They must, however, be supplied with shade as a refuge from the flies and with shelter against the extremes of heat and cold; for the English climate is apt to give us both between the beginning of May and the end of September. After September the cattle under ten months old should be housed, but older ones may be finished quite successfully on aftermath with cake. One must, however, if one is not specially favoured as regards soil and climate, be prepared against a time of drought. This involves looking ahead, or, in other words, exercising sound farming judgment, and growing something that may either be fed green to the cattle when the grass fails or made into hay or silage if the natural herbage is plentiful. Lucerne, vetches, Italian rye-grass and trefoil are all crops which suggest themselves as useful for insurance purposes in this respect. And it cannot be too prominently borne in mind that wherever the cattle may be, indoors or out of doors, they must never be allowed to go back; if they are allowed to suffer a serious check, one may say good-bye to baby-beef production. Having heard much about green-soiling being prohibited on account of labour difficulties, and about the supposed impossibility of growing baby-beef without green-soiling, it gives me special pleasure to publish the following table—a record of what happened in my own practice. The calves were very ordinary specimens indeed, they were bought promiscuously as required, on a market where the choice was notoriously poor. The animals quoted were at grass during many weeks of their lives and no green fodder was carried to them in the field. The season was, however, a growing one, and we were always ready to supplement the pasture with fodder crops in case of need.

The beeflings (Group A, in the table that follows) were *finishing* for the butcher at the same time as some older bullocks (Group B) that had not been treated all their lives with a view to baby-beef production, though they had never been subjected to the semi-starvation period that many of our ordinary stores have unfortunately to undergo.

Table showing effect of age on cost of Beef-production.

		Group A 15 months	Group B 22 75 months
Average age of bullocks:			
Daily Ration:	Linseed Cake	2 lb.	3 lb.
	Maize Gluten Feed ...	2 lb.	3 lb.
	Decorticated Cotton Cake	1 lb.	1 lb.
	Hay	10 lb.	4 lb.
	Good Straw Chaff ...	nil.	4 lb.[1]
	Roots	35 lb.	76 lb.
Average gain in live weight per month (28 days)		54·25 lb.	40·25 lb.

[1] Chaff was allowed ad lib. (*but weighed*). The cattle, however, only ate this small quantity, but they also ate some of the long straw of their litter.

It may be added this table was shown to, and discussed by, hundreds of farmers visiting the University Farm, Cambridge; these farmers were amazed at the figures, but the *relative* results are only what anyone who has thoroughly investigated the matter would expect: though neither group did particularly well. The bullocks in *both groups* were sold to a first-class firm of butchers, and the beef gave every satisfaction. The figures refer to *the last two months* of the animals' lives.

I have no experience of growing baby-beef with skim, or with separated milk alone. I confess to being doubtful about its possibility, and I am very far from satisfied that it is necessary to rely upon this food, which is very imperfect when the calf is still very young. Our neighbours on the continent have in the past supplied us with butter very cheaply, and separated milk is the accompaniment of butter-making. I very much doubt if it will ever be worth our while to try to compete with them. Candidly, I would not wish to try to introduce the conditions of life among our agricultural labourers that I have seen prevailing among the peasantry as a result of whose labours good butter was sold on our markets in England at 1s. 4d. a pound. But, even if butter-making were likely to be lucrative, the fat off 50 gallons of milk would only represent about 20 lb. of butter, and I feel sure the selling price of such butter would be well spent, in most cases, in giving the beefling a good start. If 50 gallons of milk were given whole, the "skim" from the remainder of a cow's produce would be excellent food to supplement oil cakes, once the calf were old enough to digest that form of nourishment. They reach this stage at about the age of six weeks : they begin to nibble food at about four weeks; if they are not taught, as they should be, to eat good food, they will nibble all sorts of trash at that age; at seven weeks they will eat suitable foods in adequate quantity to allow of the whole milk being gradually, if rapidly, discontinued. It is during the early days, when their digestive apparatus cannot deal properly with coarser foods, that the whole milk proves itself so valuable. Holland, which takes a large portion of her produce off her well-farmed and fertile soil as butter or cheese, spares a proportion of whole milk (*and gives nothing but whole milk*) to make veal. I suggest that, if ever it is profitable in this country to make butter, a proportion of whole milk can well be spared to grow beef in the most economical way.

In advocating the increase of production that I believe would result from the growing of beeflings in great numbers, I am bound to add that the industry cannot thrive on inferior calves. That the supply of decently-bred calves is at present altogether inadequate to ensure proper success is undeniable; so in the next chapter I propose to deal with the matter at some length.

CHAPTER VI

DUAL-PURPOSE CATTLE

ALL cows give milk and all oxen yield flesh, but the interference of domestication has led to certain varieties of cattle being suitable only for the dairyman and others only for the butcher. England, it would seem, demands an animal perfect in both respects—perhaps (even without reckoning draught, which other countries demand) we should say in three respects. For, as well as yielding a good flow of milk and providing, after feeding, a fine carcase for the butcher, a cow may also drop a calf that will make a very useful carcase of veal; though, if the same type of male offspring be allowed to grow on as steers, it does not follow that they make prime carcases of beef. If fleshing of good quality is to be found on the offspring, it must also be a feature of the parents, but many cows will themselves feed decently, at the proper time, whose calves make good veal, yet whose steers often fail in **quality**, or **rapidity of growth**, or **both**.

So perhaps the term "Dual-purpose" is somewhat too restricted. On the other hand it is doubtful whether it is possible to breed what might be called a "general-utility" animal. For there is yet another important product. Some cows give a milk so rich in fat as to be particularly well-suited for butter-making. The average deep-milking cow will give a milk whose butter-fat content is between 3 and 4 per cent. or, in some breeds, even less. Other breeds will easily average between 4 and 5 per cent. of butter-fat, if their milk is systematically analysed. Experience shows that all breeds of cows whose milk is exceptionally rich and plentiful seem to be of very poor meat-producing capacity. It might be possible to establish a breed that would be perfect for milk, butter, beef and veal, but for the moment we could be satisfied with quantity of milk, beef and veal. Unfortunately the animal that combines these good qualities (which for want of a better name we call a "Dual-purpose" animal) is very much too rare; but it can, and should, become the common stock of any country that aims at wheat *and* beef production.

It is especially necessary to any system of intensive production from the land, that includes beef-growing, that all land good enough to carry large or medium-sized cattle, should be well stocked with dual-purpose cows. The thin-fleshed cow that produces butter-making milk is undesirable, for she is worth very little herself as meat when her milking days are over. She has cost as much, or almost as much, in produce from the land to rear as has the better beef-animal, and yet when she comes to be slaughtered she is altogether inferior. Her calves are poor as veal-makers, and her male calves cannot be grown on and made into beef without great loss. Some soils in this country cannot grow well-meated cattle without the addition of excessive artificial feeding, and on such, no doubt, the thin-fleshed animal is very much more valuable than any other cow stock, and should be bred so as to produce the richest possible milk. It is folly to keep cows that are not good for some purpose or other—and at present this is too often done. These purely dairy breeds have their admirers, it is true, who advocate their being kept on all classes of land and maintain that the loss on the carcase of the thin-fleshed cow—and they do not dispute that the loss is considerable—is compensated by the extra amount of dairy produce obtained every year of the animal's life. This would be partially true if we could rely upon a long life of continuous milk-production. But the deep-milking cow, of any sort, is a delicate machine worked under very high pressure, and, like all other machines under similar circumstances, is very apt to go wrong. Udder troubles, failure to breed, and diseases of all sorts are only too often liable to end a career which is at best none too long to secure a return for the food expended in growing an exclusively dairy cow. A cow is not in a state to yield full profit till she drops her third calf; if we could rely, say, on seven periods of lactation during which she was at full profit, it would be easy to calculate which of the two animals, the dairy or the dual-purpose, was the better to keep. Unfortunately, there is never certainty; and so, wherever the land is suitable, it is always best not to trust solely to such delicate organisms as the highly-developed milk-yielder, but to have in reserve a reliable source of profit in the form of a useful carcase.

Another disadvantage of the thin-fleshed but rich milk-yielding cow is the fact that the public is not educated up to paying more for milk rich in cream than for poorer stuff. Once the prevailing ignorance about agricultural produce has been dispelled (and that will take time) improvement in this respect will depend on whether the extra butter-fat is valuable. There is no doubt that for very young infants who unhappily cannot have their own mother's milk, it is simpler, and in every way easier, to imitate the human food with very rich, rather than with moderately rich, milk. Even assuming that people generally understood the first principles of feeding very little children, and were willing and able to pay for the best material, the demand for suitable milk would still be very limited; the great bulk of the cows' liquid produce is wanted for ordinary household consumption. As to the value of milk particularly rich in fat for the feeding of growing children (after they are weaned) and for human adult consumption, it is hard to speak with certainty. There is no evidence establishing a great superiority in feeding value of the 4·5 per cent. over the 3·5 per cent. article except for feeding babies who are not fed at the breast. Many mothers do not realize the vital importance of breast-feeding and accurate knowledge about milk, generally, is very meagre. It has, however, been my business during the past twenty years to find out all I could about the management of large herds of deep-milking cows kept for supplying milk to human beings; and in this time I have enquired into the experience of hundreds of farmers carrying on this industry. Yet I have only found two cases where the milk from cows known to be very rich milk-producers was sold to a retailer for a better price than that from cows known to give very moderate quality. I have known many cases where the herd contained about 10 per cent., or less, of animals—such as the Jersey—whose milk was known to be very rich. This was done to avoid a summons to appear before the magistrates. It is a legal offence to sell milk which fails to show an analysis of 3 per cent. of butter-fat, and the milk-seller is assumed to be guilty of adulteration if his milk contains less than "3 per cent. of fatty solids" unless and until he can prove that it was in that condition when it left the cow. It is common knowledge that in our present state of cattle-breeding, the milk from herds of ordinary commercial cows will sometimes fall below this standard; it is a matter almost of everyday occurrence. This frequently leads to an unfair accusation of fraud against the owners of such stock. As it is very unpleasant to be charged with such an offence, and as it is difficult and expensive to prove a negative, some people keep a few little thin-fleshed, rich milk yielders as an insurance against the average quality of their milk falling below the legal standard.

If it is easy to make out a case against the cow who in her lifetime does nothing but supply rich milk and at her death shows a carcase which is simply skin and bone, it is very much easier to prove that the cow whose carcase is her only asset is hopeless as an aid to high production. She is chiefly useful as a means of stealing from the land without making any very high return. The following statement shows plainly the loss, from the point of view of winning human food from our limited soil, that may occur on every area of land used solely for breeding purposes. A breeding cow kept simply to produce one weaned calf will consume the produce from 2·5 acres of medium grass-land; and in very many cases, where the land is allowed to become weedy or is badly grazed or otherwise mis-managed, three acres are employed for the purpose. The calf coming off the 2·5 acres of pasture and meadow will, given fair management, be quite an average one if it weighs 400 lb. Figures before me establish the point beyond all question. Now this same area of land is shown, in the following statement, to yield the equivalent weight of calf as well as 1500 lb. of corn for human consumption, and 1600 lb. of straw for the manu-facture of four loads of farmyard manure.

I have made the following table as simple as possible, so that anyone who wishes can pick holes in the detail. For instance, the proportion of dry fodder is low. This defect can be got over by growing lucerne or vetch hay on part of the "root" breadth, or in many other ways. Probably in practice, on a farm run on these lines, it would be more convenient to replace part of the cow-stock with sheep; for these latter consume a larger proportion of "roots" to dry fodder than horned-stock. But notwithstanding any criticism of detail, the table represents a principle which, in practice, will be found to be fundamentally true, and I have estimated the yields quite low enough for safety.

Keep for Cow and Man off 2·5 acres of plough-land.

Area	Crop	Yield	Used for
·5 acres	"Seeds"	6 tons green fodder	100 days cow keep
·5 ,,	Roots, etc.	10 tons green fodder	
·5 ,,	Oats	1000 lb. corn	265 days cow keep and some farmyard manure
		1400 lb. straw	
·5 ,,	Barley	1120 lb. straw	
		1120 lb. corn	Half for cow keep, half for pork and beer for man
·5 ,,	Wheat	1000 lb. corn	Bread for man
		1400 lb. straw	100 days litter and 2·5 tons of manure

Total 2·5 acres give keep for cow, yielding 400 lb. of calf plus 1500 lb. grain for man.

Obviously, the outlay will be much greater in the case of the plough-land; the enterprise will need more capital and employ about eight times as much labour—even supposing grass-land received proper attention, the proportion of labour would still be at least five times as great. It is equally obvious that if the men working the land are to receive a decent wage, this extra outlay could not be paid for with corn at a ruinous figure. The change could not be advocated with wheat at less than 60s. and barley at 40s. a quarter, beef under 1s. a pound wholesale, and the men's wages fixed at 5s. a day. It is no use pretending that our medium-land farms can be worked intensively if the corn is to be sold at prices that merely repay operations which very much resemble theft from self-fertilized virgin soil. Even with such prices and such wages it would probably be more profitable to leave one acre out of three under grass and graze the herd of breeding cows for the greater part of the year, just bringing them into the yards for the winter to consume straw and turnips while converting all possible litter into farmyard manure. The supplementing of this store-made dung with suitable concentrated fertilizers would also be an extra charge to be added to the cost of the increased production from the plough-land.

Though the keeping of cattle for beef-production *only* on arable land makes the nation a better return from its very limited acreage, than does the practice of letting them pick up a living off grass-land, there can be no defence of the purely beef-cow under either system. The fact is that, owing to her very low output, she fails fundamentally on economic principles. A cow that gives 400 gallons of milk costs as much to maintain as one giving 800. That is to say, while in both cases a large-framed animal must be kept alive and thriving, in the one case there is a chance of a great surplus of milk, in the other there is no such chance. The difference is well illustrated by ascertaining what the gallon of milk costs in each case. It is assumed that the 400-gallon cow is the one instanced in the estimate on page 59 and that with her, as with the animal giving double the quantity, calving takes place just at the advent of the spring grass.

Cow yielding 400 gallons.

"Roots"	22,400 lb. ÷ 400 = 56·00 lb.	
Corn	1560 lb. ÷ 400 = ·3·90 lb.	Cost of 1 gallon of milk.
Dry Fodder[1]	5620 lb. ÷ 400 = 14·09 lb.	

Cow yielding 800 gallons.

"Roots"	22,400 lb. ÷ 800 = 28·00 lb.	
Corn	3000 lb. ÷ 800 = 3·75 lb.	Cost of 1 gallon of milk.
Dry Fodder	7300 lb. ÷ 800 = 9·1 lb.	

Or, putting the case in another way, it may be said that the second 400 gallons of milk only costs 1440 lb. of corn and 1680 lb. of dry fodder and no "roots." This is not strictly accurate, for the good milker would not keep in the same condition; she would lose more flesh than the "beef" animal. This, in practice, is not altogether an advantage to the farmer; sometimes it is quite the reverse. Many cows, while suckling,

---

[1] *"Seeds"* fodder reduced to hay. This cow's daily average ration will therefore be: roots 60 lb., dry fodder 15 lb. and corn 4 lb.

get so fat that they fail to breed regularly. I have known many cases where the majority of the herd had to be starved into a lean enough condition to allow of their conceiving again after rearing a calf; this means waste of time and of feeding-stuffs consumed during the delay. Again, in years of drought the cows depending upon pasture alone will be greatly troubled by flies, which are particularly bad in rainless seasons, and by the meagreness of the herbage; thus they will become so poor and thin that, through weakness, they will fail to breed.

The factor that limits intensive production of milk, meat, and corn in combination off average land is, apart from prices, the small number of good dual-purpose cattle. It is deplorable, but true, that the present supply of such animals is altogether inadequate in this country. The fact is only too well illustrated in any market or in the herds of commercial cows kept by the ordinary farmer. A very large proportion of the fat cows, and a still greater proportion of the fat bulls, on our markets, are no ornament to a civilized community. The few good specimens of exceptional worth—and there generally are some— which are offered for sale to the shambles after their breeding and milking careers are over, simply show the supineness of those who forward for sale innumerable specimens, which are as bad as can be found on any civilized market in the world. And this in a country famous for the production of high-class stock, on an island visited by breeders from all parts of the world who wish to replenish their herds with the best blood possible, among a people who for the last 150 years have shown themselves specially gifted in breeding and rearing the very best specimens of all classes of farm stock! It is strange and lamentable, but it is undoubtedly true.

If many of the beasts which England, Ireland, Scotland, and Wales present for sale to the butchers fail in respect of their beef-production, if they have skeleton in excess where the bony structure ought to be reduced, if they are so conformed that there is far too large a proportion of offal, shank, and hide, if the meat is scarce on all the most valuable cuts, and if their "touch" denotes that the whole of their flesh is too tough for consumption unless it is offered to the consumer reduced to a "potted" paste, is there evidence to show that their milking qualities in any way atone for all these failures?

The experience of inspecting a very large number of herds of all sizes in many different localities, belonging to all sorts and conditions of farmers and farms, and kept under every condition of climate and fertility of land, convinced me years ago that deep-milking were as rare as decently-fleshed cows, and that many of the worst specimens, as butchers' beasts, were no better for the pail than they were for the block. Of recent years my conviction has been upheld by milk records taken systematically in different parts of the kingdom under responsible supervision. It is true that the number of such records taken under the inspection necessary to ensure accuracy has hitherto been lamentably small, compared with what has been done in other countries. Still, it has been sufficient to enable a reliable judgment to be formed. Judging by my own experience, as well as by evidence collected by reliable milk recorders, I can say without hesitation that among our commercial cattle the number of better-class specimens of good deep-milking kine simply emphasizes the worthlessness of the poor quality beasts and shows the folly of the innumerable mediocre specimens being allowed to eat food that might be much more profitably consumed.

There is probably no better evidence of the decadence to which, owing to unfavourable economic conditions, British husbandry has sunk, than the perfection of the animals bred by good farmers holding specially good farms, and the imperfection of the stock kept by the ordinary practitioners, the ordinary tenants of medium land. These have had to face such unhappy conditions that only the men satisfied with very poor returns for their arduous labour, or those owning abundant capital, have been able to avoid a vicious circle which continually became worse as far as cattle-breeding was concerned. A consideration of these pre-war conditions with a view to future improvement will not be superfluous; in fact, it is out of the faulty practice of the past that the improved husbandry of the future must, in the ordinary course, be developed. In agriculture, which moves slowly, a sudden and drastic change that will instantly transform the evil of one year into the perfection of the next is impossible; for a decade, which represents rather less than three generations of stock-rearing, is the least period by which improvement in farm-stock can be measured.

Twenty-five years' observation of farming before the war showed me that apart from robbing the land by running store-stock upon it, there were, with certain exceptions, only two ways of getting a decent living from the cattle industry.

The first was growing and selling milk (generally to those living in towns) for consumption in its primitive form. The conditions under which this could be carried on profitably were by no means universal and not always continuous. To ensure success it was essential for the farmer to be favourably situated as regards transport, to have a special water-supply connected with his holding, and to be able to secure milkers. These conditions were by no means universal, but at a time when other farming operations offered but little promise of a decent reward for enterprise, they were common enough to encourage a supply that was often at least as great as the demand. This led to a frequent fall in the value of milk, which was practically the only monopoly the British and Irish farmer could produce. The price, indeed, often fell so low that only by rigid economy, not to say parsimony, could it be produced with any hope of profit; in fact, the prices prevailing in some years were such as to make one marvel how the farmers who produced it kept going at all. One very easy form of economy was to buy a very inferior bull to mate with the cows. This was a very insidious form of false economy, for many of the milk-selling farmers did no rearing at all. The majority, in fact, sold off their calves at from three to seven days old. They looked upon breeding as a necessary evil; had they been able to keep the cows in profit without the trouble of producing calves, it would have suited them well. That the heifer calves had to grow on and make the future milkers did not immediately concern them, and so they were responsible for the gradual deterioration of a very large proportion of the cow-stock of the country. This was a short-sighted and disastrous policy, not only from the national point of view, but from their own. For such a large proportion of all our cattle was—and still is—brought into existence in the herds on the milk-selling farms, that the man who bred them and sold them as babies was often obliged to buy them back when they had reached maturity, and thus replace his own worn-out milkers. In fact, so great had the proportion of ordinary and inferior cows become, that all really good specimens made a price which only those very favourably situated for selling milk could pay. Thus a new evil arose, as many of the cow-keepers who were able to buy the best milkers carrying good flesh did not breed from them; it suited them better to milk through one period of lactation and then sell them to the butcher, the processes of feeding for milk and for fat meat going on concurrently. This system, called "milking out fat," has the worst possible effect on the quality of our cow-stock; the only excuse for it is that the milk-sellers who practise this are saved the risk of disease—tuberculosis and contagious abortion—which undoubtedly exist to a very alarming extent among our dairy cows.

On the average farm the evil generally went on from bad to worse. The calf resulting from the use of a bad bull on one farm grew into an inferior cow on a distant holding, for it was seldom economical to rear her on the milk-seller's farm; she was then sold into some cow-shed adjacent to the railway or residential district and was mated with another inferior bull. The result was a calf of still poorer quality and so the vicious process was continued. It is true, though no excuse can be made for it, that our general **want of method** in distribution and **co-operative buying and selling** gave some small justification for such short-sighted policy on the part of those who kept cows only to give milk. Dealers buying young calves paid the same price—and that, of course, the lowest possible—for one inheriting good milking and fleshing qualities as for one sired by a bull with no good qualities to transmit. Thus a farmer who paid £35 for a good bull to use in his herd instead of buying something unspeakably bad for £10 or £12, got no immediate return for his outlay; the utmost, in fact, he was ever likely to get was an occasional good animal which he kept to rear himself. The number of those so kept, however, was often so small that it made the cost of each one, owing to the amount spent on the purchase of the sire, too expensive to be profitable. A bull costing £100 may be cheap to the breeder who rears, as he may well do, a hundred of his offspring; but a bull costing £35 may be a very poor investment, if not more than two or three heifers of his getting are reared in each of the years during which he stands at service among a lot of milch-cows whose offspring go to market and make no more than the poorest mongrel calf. The wealth lost to the nation through the failure to encourage milk-sellers to breed good calves must be enormous; but, great as the amount must be, the loss on all youngsters sold on our markets, due to our lack of method, has been even greater. The question of improvement in this respect will be discussed in another chapter.

The second profitable system of cow-keeping was cheese-making, but this, too, required special conditions: suitable grass-land not too highly rented, a good and permanent supply of water, and, as before, a supply of milkers. But cheese, unlike milk, had no monopoly protection, and so had less security against the ruinous competition of dumped goods. It must be admitted that this competition often came from farmers who, by improved methods of manufacture and marketing and by breeding only suitable cattle, were able to under-sell our own agriculturists. Even with their improved methods (and only insular prejudice will deny the superiority of their general management) it is doubtful whether the price at which their cheese was sold was enough to secure the workmen of the producers a living wage. Visits to the best cheese districts on the continent revealed a manner of life among the labourers which showed the conditions of our best-housed men in a very favourable light (not that those prevailing in our own cheese-making districts, though generally better than those in purely corn-raising areas, were anything to boast of); for instance, it was quite common for a cowman's personal accommodation to be simply a sleeping-box in the wall of the cow-house!

The competition the cheese-makers had to face was not always such as to make their industry unprofitable; it was only occasionally that an excessive supply of foreign or colonial produce ran prices down to a figure that brought the returns very close to the cost of production, and it was only in very bad years that prices actually fell below the cost of production. But the risk of this happening produced a feeling of insecurity, which in the long run must always tell against enterprise and large-minded action. It was, no doubt, the prevalence of this feeling, combined with other reasons, which prevented a systematic improvement of cattle among the herds of the cheese-makers.

The sires used in the cheese-making herds were not such terrible specimens as were, and are, commonly used by milk-sellers, but there was, nevertheless, much room for improvement. Here and there a breeder had the very finest stock the world could produce, but, generally, the cheese-making farmer was content with very mediocre cattle indeed. Nearly all those occupied in this industry reared their own stock. Only a small percentage of heifer calves, however, were kept to fill, in due course, the gaps made by disease, accident, and old age in the cow-houses; the great majority of calves were eagerly sold soon after they were dropped. The more careful farmers kept heifers from what they believed to be their best milkers, but as the milk given by individual cows was hardly ever recorded, improvement in this respect was uncertain. But it was only the exceptional man who made a practice of securing the services of a sire known to be of deep-milking blood on the female side of his ancestry, and showing the quality of begetting well-fleshed descendants. As with the milk-selling farmer, so here the evils of an **unorganized** calf-trade arise largely from the unscientific, unpractical, and disadvantageous system of breeding in vogue. A few days after birth, the great majority of calves were bought by one man who dispatched them to another in a different part of the country; thus, I have seen shiploads on their way from Somersetshire to Ireland. They were then sold to a rearer, who had no means of judging their quality except by eye and, after a long journey, most calves look poor things indeed. Even if a rearer happened to buy a specially good "bunch" of calves, he had no means of finding out anything about their origin, and so could not be sure of securing another similar lot, however willing he might be to pay such a price as would repay the cheese-maker for laying out money on a really good bull. But, usually, the calves by a good bull were not sent about the country for promiscuous sale. The careful and upright dealers would secure them for their own regular customers, and so the great bulk of the British rearers grew accustomed to mediocre stock.

It will easily be seen that a cow that will give 800 gallons of milk yearly, or (allowing for the reduced flow after the first two calvings) an average of 750 gallons throughout a milking life of seven years, and is also quite a good beef producer, is superior to what nature would produce unaided by skilful selection of parentage. To populate a whole countryside with such animals requires systematic effort and co-operation intelligently directed. It is the absolute failure of comprehensive collective action to secure for the home industry a supply of animals best suited to their requirements that has led to our country being covered with cattle very much below the standard just suggested. In the absence of State aid, and of systematic assistance by those who ought to have taken the lead during the bad farming period between 1870 and 1900, nearly all the magnificent efforts of our countrymen, born live-stock improvers as they are, were concentrated on satisfying the demands of those overseas customers who offered higher prices than were obtainable from the much poorer English tenant-farmer. As the overseas customer wanted cattle for purposes of stealing from the soil, and as the home agriculturist was in many cases struggling to farm in competition with those stolen goods, their interests, so far from being always identical, were in fact more often altogether antagonistic. In the next chapter the evils of these conditions will be considered in some detail, for in them lies the chief cause of our having so many inferior commercial herds of cattle.

CHAPTER VII

PEDIGREE BREEDING

THE expression "Pedigree" as it is used in the cattle world has come to include a great deal. It implies that an animal with a pedigree has been bred from ancestors who for several generations have been selected as breeding-stock for special reasons, and with certain definite objects in view, and that, consequently, it has itself inherited and is capable of transmitting these qualities to its offspring; further, it implies that the animal's name, together with those of a certain number of its forebears, have been registered in a recognized record or official herd-book. The early improvers of the modern breeds of British cattle died without making any official register, and the methods of creation of our present breeds of cattle may be said to have begun with the work of Robert Bakewell (1725–1795); but it was not till 1822 that the first official register of cattle—the Coates's Herd Book—was published. The careful mating of parents and the selection of their offspring with a view to their imparting certain valuable qualities had, however, been practised from the first; this principle, *it is generally believed*, constituted the whole essence of the practice of Bakewell in accomplishing his wonderful improvement.

Caution is necessary in speaking of the methods of Bakewell and the other pioneers, for we have little authoritative record of how they carried on their important operations. The professional historian has not deigned to spend much time in collecting information for the mere enthusiast about such humble creatures as cows, calves, and bulls. The fragments of knowledge we possess have been picked up here and there, and either gathered together by an agriculturalist untrained in methods of historical research or, more frequently, left as scattered allusions in the works of historians who cared little about agriculture and still less about cattle. This is very strange, for historians realize that the improvement in agricultural practice of the eighteenth century played a large part in the success of the early nineteenth century, a success without which this country might have fallen very low instead of rising to eminence, as she did, at the opening of the twentieth century. An exceptional instance of foresight on the part of a statesman[1] supplied money in 1910 for special advanced training in agriculture and for research in all matters connected with it. Nothing, however, has been spent on the subject, though it is admitted that the breeding of improved cattle played a very large part in the improved agriculture which helped us to fight the Napoleonic wars[2]. The authorities apparently decided that practice was altogether unworthy of financial help and therefore left the husbandman without any inducement to systematize his knowledge. There is, however, no one in greater need of systematic training in the practice of his profession than the average farmer, and yet none of this money was made available in the interests of work that might lead to empirical methods becoming more accurate. Agricultural science, to which much of the money granted since 1909 has been devoted, is bound to be handicapped if it is not combined with a well thought-out practice; for, after all, the teaching of chemistry, botany and the like are of little use to the farmer if not superimposed upon good husbandry. It seems to have been forgotten that it is practice, the actual farming operations, that wins food from the land for man. There was disappointment, discouragement, and mortification amongst those who wished to help farmers and stockbreeders by spreading knowledge of husbandry born of critical and systematic research, at having no funds to meet the expenses of the thorough and continuous investigations without which no teaching can be really useful. Until a great and continuous effort is made to systematize the work on farms, our husbandmen (like their cattle) will continue to have among them a few of the best, and many of the worst, types of their profession.

If the history of the improvement of British breeds of cattle

[1] The Right Hon. D. Lloyd George, the present Prime Minister, in his famous Budget of 1909 put aside £2,000,000 for the Development of Agriculture.

[2] *Napoleonic Studies*, by J. Holland Rose, G. Bell and Son, pp. 195, 196 (and footnote, p. 196).

is very scanty, the knowledge of existing conditions is still less diffused among our population. I once addressed about 180 men and women who were teachers in our rural elementary schools. The gathering expressed some dissent at certain remarks I made about the teaching in our village schools being given with little or no sympathy towards agriculture. When my turn came to reply to the various criticisms that had been made, I asked a question myself. It referred to their teaching of history. I invited all those who had included anything about the work of Robert Bakewell in their history lessons to hold up their hands. Only three hands were held up! More than 175 out of 180 rural teachers had to confess they had never mentioned his name to their pupils. If this was the state of affairs, about 1910, among the instructors of children about to spend their lives among cattle and sheep, it is easy to imagine that the boys and girls at our secondary and public schools had no better enlightenment about rural history. It is probably true to say that the vast majority of Englishmen of the last generation lived and died without ever having heard of Bakewell, Townsend or Tull, the three men who, between 1720 and 1780, made our agriculture the best example to all husbandmen in the world, and the first of whom has brought millions of money from overseas into the national exchequer.

I lay great stress upon the unfortunate ignorance that has prevailed among all classes about the work of these great men, for this ignorance was largely responsible for the terrible state of affairs prevailing in 1914, that allowed our enemy to threaten us with starvation while we were living amid fertile, but almost uncultivated fields. The exceptions, the cultivated fields, showed by their very excellence how much the strictly limited amount of our national land might have been made to produce by universally good farming. In times when we least wished to let the enemy hamper our military effort, fighting material was necessarily sacrificed so that the shameful neglect of the past might be remedied with a slowness born of hurry. If the educationalists of the past two generations had deemed agriculture worthy of their attention, how much loss of blood and treasure through U-boats might they not have saved?

Though Robert Bakewell was the pioneer in breeding operations among cattle, his great influence lives through his disciples rather than through his cattle, or even his sheep. We know that he worked with a breed known as the "Longhorn" and that in spite of the improvement effected in that variety by his skill, enterprise and pertinacity, it never became very widely distributed. The brothers Collings certainly under the indirect influence of, and probably directly inspired by, Bakewell, fixed upon other stock than his for improvement. It was the descendants of these animals, now called by us the "Shorthorn," that spread everywhere. There is good evidence to show that the animals with which the Collings brothers worked had themselves been much improved by others before them. It appears from existing records that there had always been good native cattle belonging to the agriculturists who lived in the North-Eastern counties of England, the home of the Collings family. These native cattle had no doubt been crossed with good bulls imported from neighbouring continental countries. It is possible that the breeding material upon which the Collings began to work was itself better than that with which Bakewell started. Moreover, the brothers had as collaborators a very fine body of agriculturists working during their lifetime on the creation of the same breed; and it would appear that their success, over other breeders, was as much due to their powers of self-advertisement as to their skill as breeders and rearers. Great as this skill was, their powers of advertisement brought their names into much greater prominence than those of any of their contemporaries. The history[1] of their work is full of interest, but here there is space only for an analysis of results, which, it is hoped, may be of use in the immediate future.

Whatever were the qualities of the cattle with which Robert and George Collings began their breeding operations, they were at the end of their careers turning out beasts eminently suitable for stocking the farms of the greater part of England, Scotland

[1] *History of Shorthorn Cattle,* by James Sinclair, Vinton and Co., 1907. In this work the author has collected from various sources a valuable amount of information concerning the doings of the early improvers as well as much other useful information which brings the story of the breed up to date of publication.

and Ireland—there is not as full a record of their going into Wales. The customers whose requirements they had in view were the yeoman and the tenant-farmer or his landlord. For, in those days, the landlord and farmer were not thinking of the foreigner's cheque; the landlord kept and bred sires with the sole object of improving the animals that were useful in increasing the produce of his own land or that held by his tenants.

It is now time to review the good qualities and the imperfections of this stock which has become universally recognized for its great merit, at any rate wherever the English tongue is spoken or the supply of beef for English markets is considered; even though the rearing-ground may be thousands of miles from our shores.

The Shorthorns brought into being by the brothers Collings and their fellow-breeders had the following characteristics as compared with the foundation stock-animals from which they sprang. The bony frame, or skeletal structure, was reduced to the smallest limits consistent with usefulness. All excess of bone not needed to support the body weight, to admit of an adequate covering of muscle and its attachments (sinews and such tissue), or to protect the delicate internal organs was, by careful selection of sires and dams, bred out of the cattle in their herds. The shape of the frame was thus greatly improved. The depth of carcase was much increased in proportion to the total height of the animal, length, as well as weight, of limb not being required for beasts that were not wanted for labour; the width, too, of frame, or body, was wider in these "Improved Shorthorns." This applied especially to the fore-end of the animals, for there a wide-spreading conformation of the ribs was combined with a great depth of frame to give ample room for development of heart and lungs. These vital organs of circulation and respiration, not being cramped for space, admit of the development and full activity of the necessary bodily functions (growth, digestion and so on) with greater profit to man. Besides reduction in size, and improvement of the shape of frame, a great development of meat, or of "*thick flesh*" as it is called, was also effected. Not only was there an increase in the proportional amount of meat to bone and in the proportion of valuable "cuts" to second- and third-class species of "joints," but all the edible parts, the lean meat, or muscle, and fat were greatly improved in quality. This question of "quality" will, on account of its economic importance, be referred to again. To all these improvements the early breeders of Shorthorns added another very valuable, and, as regards their early-owned cattle, quite distinctive attribute of usefulness, namely, the faculty of rapid growth or early maturity. Bakewell's development of this faculty in his Longhorn cattle had seemed miraculous when compared with the normal growth of oxen before his time; the Shorthorn-breeders were not one degree behind the pioneer in this respect.

What happened to the milking capacities of the breed, while this transformation in fleshing quality and early maturity was going on, has never been quite clear to me. It seems certain that the bulk of the Teeswater or Holderness herds, among the individual members of which the improvers found their original breeding material, were very famous for their milking qualities. There are many testimonials, independent of any breeders' influence and quite separate in their source, in the *Annals of Agriculture*[1] and other literature of the time giving evidence of immense yields of milk from the "foundation" cattle, and it is probable that their successors, the improved Shorthorn, did not milk so well. On the other hand, there is no evidence whatever, quite the contrary[2], that the milking qualities had almost disappeared; and to suggest that they had vanished to the same degree as was the case some hundred years after the Collings brothers were breeding seems to me ridiculous. The fact probably was that no Shorthorn-breeder paid very particular attention to the amount of milk his cattle gave. He seemed, however, to assume that his cows would, by virtue of their inherited characteristic, give an average of about 800 gallons a year without special attention being paid to the selection of stock for their pail-filling qualities. It was in beef that his

---

[1] *Annals of Agriculture and other Useful Arts*, 1790 to 1797, by Arthur Young, Esq., F.R.S.

[2] Sinclair, *loc. cit.*

original stock failed. He wished to improve the quantity, the quality, and the rate of production of meat among his deep-milking cattle; to found, in other words, a dual-purpose breed, and there is ample evidence that, although no particular care was given to deep-milking qualities, this object was completely attained. But, before developing this theme, it is interesting to consider some other factors.

While all these "utility" points were being developed, an equally important change was effected in some minor matters. These details are sometimes scornfully spoken of as "fancy points," that is, points of colour, shape and carriage of head, carriage and texture of horn, and even absence of black pigmentation in both horn and muzzle—all points which in themselves seem utterly worthless and have consequently been ridiculed even by the most intelligent and broad-minded friends of agriculture. Such critics held that the shape of head in no way affected the value of a cow or bull. "What," they asked, "is the value of width between the eyes, of delicate chiselling-away of the bony outline of the profile, of the 'dished-face[1],' of the deep, well-turned jaw with no superfluous skin hiding its conformation and musculature, or of the waxy, flat, cream-coloured horn well placed on the head and showing no black in the tip?" The answer of the practical man in the past was extremely vague. He would reply that such points made his animals sell, or that their absence brought the price down, or that Mr So-and-So was, at any rate, of this opinion, and so on. Occasionally he could give a more reasoned answer: he might explain that such and such an imperfection denoted impurity of blood, that it showed that the progenitor, belonging perhaps to a remote generation, had been of another race, possibly of an unimproved breed. But every answer indicated vagueness and looseness of knowledge. One practitioner of repute would declare that experience and observation had taught him to value a particular point in an animal, while others, equally reputable as "practical men," had made no such observation.

---

[1] This feature, when combined with a fashionable pedigree and other "breed points," added £1000 or more to the value of the animal about 50 years ago!

The indefiniteness and inaccuracy of knowledge of present-day breeders and exhibitors give no clue as to whether the early improvers had good reason for their insistence upon such detail, or whether the existence of such points was at first merely accidental.

There seems, however, to be some hope that the scientific work of the present day will throw light on such points. The discovery, for instance, of certain laws relating to inheritance by Mendel (1822–1884) gives great promise of enlightenment. Mendel's successors have discovered that certain characters are associated with others, which may lie dormant, or inactive, or hidden in a living body, until the union of two parents, each carrying the component factors, causes them to "couple" and come into full view. The existence of these signs of hidden qualities, or factors, carried, but not displayed, by creatures themselves, but capable of being transmitted to offspring in an active form, seems a reasonable explanation of the great stress laid on points that may appear, at first sight, to be purely fanciful. For instance, I have noticed that the dark or "smutty" colour of the nose often seems to be associated with an undesirable texture of coat. But no "*Zootechnicien*" (to use a French word in the absence of an English one) has at present sufficient knowledge about the foundation of such points. When Professor Biffen began his successful career as a plant-breeder he had, as a botanist, a full acquaintance with the structure, and many of the functions, of the body of the wheat-organism, and he was able to obtain information of paramount importance about the composition of grain from the chemist and physicist. If the results of scientific investigation into the breeding of farm-stock are to bear full fruit, the "Zootechnicien" must remedy his ignorance which at present must be admitted to be abysmal. He must seek the assistance of the physiologist, the anatomist, the zoologist, and engraft their teaching on to the empirical knowledge of the breeder, feeder, and butcher. Without a fundamental examination and study of the structures and functions which go to build up a useful type of ox, it is hopeless to try to do more than has been done by the wonderfully successful "rule-of-thumb" practitioner in

Great Britain, whose chief weapon in his great contest with nature has been his wonderful, intuitive judgment guided by experience. On the other hand, the practising farmer must be convinced that his forebears obtained results despite, *and not because of*, the absence of scientific assistance.

To return to the work of the early Shorthorn improvers, there was no point of detail in their stock on which they seem to have set more value than that known as "touch." This word covers a great deal in connexion with the hide, the tissues lying under it, and the coat covering it. The foundation stock are reputed to have had thin and hard skins, inelastic and tightly stretched over a mass of tissue which had no feeling of springiness when examined by hand. The Collings brothers made a special point of securing animals having a totally different "touch." The frames of the animals they sold or let at what were then fabulous prices were covered with hides moderately thick, very ample, and very elastic. The flesh under the frame was required to be firm yet resilient, and nowhere to adhere closely to the animal's frame. Under pressure from the fingers the hide had to run freely and pleasantly over the whole frame, more especially over those parts where the covering between it and the bone was slightest. Failure in any of these "points" disqualified the animal for use as a sire in the herd.

So much for the hide itself. Now as to the hair. The authorities insisted that this should be very abundant and very soft, and they seem to have preferred it to be specially curly all over and to show a fine fringe on the ears, and a silky tassel at the end of the tail. The type of hair required is often described as "mossy," the very opposite of being harsh, or wire-like, to the touch. This matter of quality of hair seems to have excited the special ridicule of "systematic" workers, and I have myself known scientific authorities raise a laugh at the expense of the breeder who laid stress upon, and would pay high prices for, breeding animals with *mossy* coats. Yet some recent work on the effect of *skin-temperatures*[1] by Professor

[1] Professor Wood read a paper before the British Association at Sydney in August, 1914, on this subject. The work unfortunately has had to be discontinued on account of the war.

T. B. Wood shows how important a matter the loss of heat from the surface of an animal's body may be in relation to the return made to the feeder for provender and other nourishment supplied to it. If this is so, surely the instinct of the breeders who insisted upon a certain type of coat to secure the best result should not be ridiculed; and in the future intelligent observation which leads to a conclusion, by experience in the field, that some "point" is necessary should be deemed as worthy of consideration as any work done within the walls of the laboratory. It should not be beneath the dignity of the man of science to analyse systematically the deductions and observations of the practical breeder. Granted that many "points" will be found to be fanciful, many requirements to rest on insufficient observation, many conclusions to be based on the prejudice of ignorant, though strong-minded individuals, but all this simply shows the necessity of systematic, thorough and large-minded research—call it science or what you will. It is indeed strange that England (the "Stud Farm of the World"), which has in Rothamsted the world's pioneer experimental station for the study of plants, manures and soils, should be without any station whatever for the study of the reproduction, structure and functions of farm live-stock in 1918. Rothamsted was founded by Lawes and Gilbert in 1843.

The breeding of pedigree Shorthorns was carried on through the nineteenth century by innumerable enthusiasts in all parts of the Old World and the New, but between 1810 and 1885 there are three names that stand out more prominently than others as having contributed towards the triumphant increase and wide distribution of the breed. These names are Bates, Booth and Cruickshank. It is interesting, and very instructive, to note that all these breeders looked upon the Shorthorn as a dual-purpose animal. Bates seems to have paid more attention than the others to good milking qualities, but neither Booth nor Cruickshank[1] ignored or neglected them. As regards the ordinary stock of the country in 1908, according to the census taken by the Board of Agriculture in that year, no less than five-eighths of the cattle in England were of the Shorthorn type.

[1] Sinclair, *loc. cit.*, p. 102 *et seq.* and p. 738.

Unfortunately for the tenant-farmers of Great Britain and Ireland in particular and for the whole agriculture of the country in general, a change came over the objects which the pedigree breeders had in view during the last quarter of the past century. This change was coincident with very bad farming days in England, when miserable prices for all produce made it difficult for farmers to find money to spare for the purchase of specially good breeding stock; it was also coincident with the rise of a great trade with foreign buyers of pedigree stock who came to our shores with full purses and objects of their own. These objects require very careful study, for they materially affect the whole situation. First, they aimed at supplying our own markets with beef from cattle directly descended from our own exported pedigree stock. Secondly, they wanted an animal well fitted to take care of itself. They had practically unlimited land in a suitable climate, the cost being so trifling that any labour expended upon the care of their cattle was practically their only annual expenditure, the interest and capital outlay upon breeding stock not being included in the cost of rearing. Consequently, so far from demanding, they even objected to deep-milking cattle. So long as a cow could give her calf a good start it suited their purpose that their breeding stock should not transmit high milking-yield qualities. The udders of deep-milking cows are always much more liable to mishap than those of animals giving little. More especially is this the case with cows that are suckling, for in the early days the very young calf cannot consume all the contents of the gland, though as it grows older it will consume more than it can properly digest. These facts may lead to dire consequences both to mother and offspring, and both, therefore, need constant supervision and restraint: the cow may require hand-milking for a time and the calf may have to be separated. This requires expenditure both on labour and on the construction of enclosures and so working expenses are increased without giving any additional return whatever under the **conditions of estancia cattle-keeping.**

Spurred by these demands, our breeders made efforts, and very successful efforts, to produce Shorthorns and other cattle showing the finest qualities of beef-production and possessing truly wonderful constitutions. The best evidence of this constitution is seen in the hideous state of obesity to which breeding stock were brought in order to satisfy the demand for fleshing qualities by overseas purchasers. In a truly pathological state of corpulence bulls and show-cows and heifers were purchased, transported thousands of miles on shipboard, submitted to a long period of quarantine, after which they reproduced themselves in a climate differing totally from that in which they had been reared. Milk production was absolutely neglected. A thick-fleshed, good-coloured, well-haired cow showing constitution was used for breeding, and her descendants reproduced themselves, even though, after being fattened for sale or show, she did not give enough milk to rear her own calf.

No one can blame the breeders, who were working for their living, for catering for a foreign market which required animals quite unsuitable for our own. Indeed, very many of our own most successful farmers, men who were most useful through their influence in local government, owed their financial success almost entirely to the big prices they obtained from foreigners for this class of Shorthorn stock yielding a splendid carcase, but no milk. No one can wish for a moment to deprecate the catering for such animals at our leading exhibitions, since the money they brought into the country was too precious in the bad agricultural days for such stock to be neglected at shows.

On the other hand, it was not edifying to see the landed proprietors take up these purely beef-breeds, so suitable for the foreigner, at the expense of the dual-purpose animal suitable for their own clients, the tenant-farmers of England. Landowners should have had sufficient knowledge of farming conditions (no well-informed person will for a moment deny their good will) to prevent our leading agricultural society, which they largely controlled, from devoting more prize-money to stock suitable for the Argentine than to animals suitable for the struggling farmers of England. Yet a study of the show catalogues of one very influential body shows that from the year 1911 to 1914 £800 was given in prize-money for Shorthorns, the purely beef animal, and that during the same period

£316 only was offered for pedigree dairy Shorthorns, the dual-purpose breed.

A few words about the classes provided of late years for the "Dairy-Shorthorns" may usefully be added. By about 1899 it became evident that the milking qualities of the breed were in serious danger of being lost altogether among pedigree animals. So in 1905 some breeders started the Dairy Shorthorn (Coates's Herd Book) Association with the following object: "The aim and object of this Association is to promote the breeding of the pure-bred Dairy Shorthorn."

The name of the Association is a little misleading, for its real object was to re-establish the dual-purpose capacity of the breed. This the Dairy Shorthorn Association proved by its instructions to the judges it nominated. There is plenty of evidence to show that it was these qualities of the early Short-horns that had made them invaluable to English husbandmen. For instance, the average farmer would have nothing to do with the bulls of the purely beef type. At the public auction-sales of such cattle he would let *wholesale* butchers buy truck-loads of young bulls, sires that were not good enough for the foreign market. The expenses of exportation were great, as insurance was very high[1], and so only first-rate animals, capable of becoming the sires of bulls, were exported. The colour factor, again, led to a great number of bulls being available for the home-market. Although the exporter was buying sires to beget the breeding stock that was to reproduce purely commercial beef animals, he was most particular about the pigmentation of the hair. I remember hearing a partner of the greatest auctioneer of my time, in conversation at the Royal Show at Park Royal, telling how his colleague had had the dressed skin of a famous Shorthorn bull sent to him all the way from Chili so that he might execute a commission to export animals of *exactly the same colour*. It is common knowledge among Short-

---

[1] This was due to the risk of rejection after failing to pass the tuberculin test, for, as our ports were closed from 1896 against the entry of breeding cattle, an animal failing to pass at the place of landing had to be slaughtered; it was not allowed to return to England. For some years before this date farm-stock were not allowed to land, except for immediate slaughter, but in this year Parliament finally excluded cattle.

horn men that white legs might reduce the value of a bull by from
£50 to £200, or more, on account of the prejudice against such
markings existing in the minds of the South American buyers.
Again, these customers were most fastidious about the pedigree, particularly on the female side, of the animals they took
away. Thus there were, all told, many quite superior beasts
rejected by the foreigner for lack of colour, breeding, or quality,
and left behind for the farmer to buy. Yet he would seldom
have one of them even at the price of one half guinea more than
the butcher was willing to pay for an animal to slaughter and
to sell wholesale as bull beef.

That the farmer did not believe in the purely beef animal is
shown, again, by the patriotic, but unenlightened, efforts of
some landed proprietors to improve the breeding stock of the
country on their own home-farms. Perhaps it is more accurate
to say that the effort was generally made by the land-agent—one
of many instances of the landlord's neglect of his own particular
business. Proprietors, through their agents, have often been
known to secure a very good beef Shorthorn sire for the service
of the cows belonging to their tenants; but these animals were
as often almost entirely neglected by the men for whose special
benefit they were purchased, frequently at very considerable cost.

The subject might easily be made tedious if all the evidence
of the unsuitability of the beef Shorthorn for general farming
conditions were collected, but one last point may be mentioned,
namely that the belief became almost universal that the Shorthorn was a non-milking breed. I have heard this maintained
by agricultural authorities of all grades in this country; I have
known it taught at first-class agricultural education institutes
on the continent.

For their successful efforts in dispelling false ideas about
the non-milk yielding qualities of all Shorthorns, the agriculturists of England owe the Dairy Shorthorn Association
a very deep debt of gratitude. The difficulties they had to face
are clearly shown by a study of the list given here of the number
of entries in the classes for Shorthorns of both descriptions of
stock at the shows of the Royal Agricultural Society of England
for the years 1906 to 1910.

Table of Exhibits of Shorthorns and Dairy Shorthorns
(Coates's Herd Book) *at the R.A.S.E. Show for the years from*
1906 *to* 1910.

| Date | SHORTHORNS | | DAIRY SHORTHORNS | |
	Cows and Heifers	Bulls	Cows and Heifers	Bulls
1906	107	139	16	0
1907	90	170	21	0
1908	114	159	21	0
1909	140	180	30	0
1910	99	141	46	13[1]

In the work of restoring the Shorthorn to its high place among dual-purpose cattle, another danger arose—a tendency to look upon the value of the stock as solely determined by their power as milking-machines. A bag of bones, or "a toast-rack carrying an udder," giving a prodigious yield would, if some breeders and judges had had their way, have been encouraged by the award of prizes, whereas what was really needed was a cow giving a profitable yield of milk together with an excellent carcase for the shambles after her best milking days were done. The thanks of all lovers of the rural industry are due to all whose influence arrested this tendency. Of the men with a large interest at stake in "England's greatest industry" there were lamentably few who took any interest in the foundation of the Dairy Shorthorn Association[2]; among those who both took an interest in it and helped to guide its efforts for the encouragement of dual-purpose cattle useful to English farming, the numbers of landlords and agents are still smaller. But a tribute is due to Mr C. R. W. Adeane[3] and his agent Mr F. N. Webb,

[1] In 1910 for the first time prizes were given for bulls known to be carrying dairy qualities through the milk-yielding capacity of their dams.

[2] A study of the exhibitors' names in the classes from which my table is compiled will confirm this.

[3] Messrs Adeane and Webb were Cambridgeshire men, both natives of a district not supposed to be famous for its dairy cows. Yet their efforts, backed up by the work of the first Live-Stock Officer (Mr W. P. Crosland), appointed in 1914 under the Development Scheme, have led to amazing results which are to be seen in the first *Register of Good Dairy Cows* published in 1918 by the Board of Agriculture and Fisheries. Cambridgeshire has a wonderful record, the largest *proportion* of cows of any county are on the *Register*, and, in spite of its small area, it has the second largest *total* of cows registered.

who were both untiring and undaunted in the work; for much of the very considerable improvement now prevailing—extraordinary improvement when the years 1917 and 1918 are contrasted with 1907 and 1908—is due to their efforts. A good foundation has now been laid for the wide development of the Shorthorn in the full usefulness of its original dual-purpose capacity.

The work of the pedigree Shorthorn improver was gradually taken up by the breeders of other varieties of beef and dual-purpose cattle. The special qualifications of these will be described in other chapters, but there is one point of considerable interest to which attention may here be drawn.

A study of the markets of the last two decades will show that for export the Shorthorn has generally been sold to foreign buyers at much higher figures than any other breed. It is probably safe to estimate that, if the best breeding stock of other breeds has made an average price of £100 per head, the same class of stud animal of the Shorthorns' Herd Book has made £400. Not only were the prices higher, but the numbers of Shorthorns exported were very much greater. With slight variations this has gone on for many years. So great have been the difference in price and the numbers exported, that for many years I have wondered why purchasers from all parts of the world who merely wanted to start herds for the reproduction of beef should have shown this marked preference. Other breeds are as hardy, show as great power of rapid growth, are equally good graziers of somewhat inferior pasture, make equally saleable carcases of meat when fat or finished, yet the preference for Shorthorns continued. Fashion, the want of enterprise on the part of the breeders of other stock, the encouragement given by leading British Agricultural Societies are all given as explanations of this very marked preference, but they are not satisfactory. Apart from the fact that the foreign commission agent is quite smart enough to see through all such artificial factors for himself, there is another reason for not accepting these explanations. The strange preference has existed so long, so many specimens of other breeds have been exported, and so much capital has been expended upon the industry that one is driven to the conclusion that the preference is based upon some sound economic reason.

I have looked at the question from many points of view; I have discussed it with nearly every class concerned in the cattle industry, and I have had the good fortune to meet very many successful commission agents, breeders, feeders, and meat salesmen. I have, however, still to find **proof** of what constitutes the excellence that gives the Shorthorn this great predominance over all other breeds. I have formed a theory of my own, based upon careful observation, but I have not hitherto had an opportunity of putting it to the test. Research which is to establish a working certainty sound enough to be applied to scientific breeding is a costly matter and demands a man's whole time. Hitherto the money and therefore the time has been wanting. Until some rich person, or body of persons, supplies the necessary funds, research is impossible. It should, however, be obvious that until such characteristics as those of the Shorthorn are discovered and recognized, no advance in the practice of breeding is likely to take place; for the scientific discoveries relating to the laws which govern inheritance will not, until then, be made available to the pedigree breeder.

The last considerations, though they greatly concern pedigree cattle, are so general to the industry that they must be treated in the next chapter. I would conclude this part of my subject by an appeal to the breeders and those who influence the breeders of pedigree cattle *to give to the home-market* the same attention that they have paid to the overseas demand. No one who has watched, as I have done, the overseas trade for a quarter of a century can fail to appreciate the skill with which the customers' demands have been met. No one with any love of our countryside could wish that that trade should be neglected in the smallest degree. But it is wrong, as has so often been done, to leave our tenant-farmers either to select the misfits from the foreign market or to fall back upon mongrels that are a disgrace to any country. So great an evil do I consider this latter practice that I am distressed whenever I hear of one of our home breeds of cattle being "taken up" by the exporter. I am still Scotsman enough to realize the financial benefit such a demand brings to a considerable number of those owning the particular variety that becomes fashionable among foreign buyers. Yet the knowledge that *the whole body of breeders* may turn their endeavour to producing stock *suitable* for conditions across the water, though they be *unsuitable* for the locality where the breed is doing useful work at home, makes me wish that the·overseas agent had stayed away. The pity of it is that this state of affairs need not exist. Systematic study of the conditions under which cattle are to be kept, and research into the qualifications of the animals suitable for those conditions would prove that there is plenty of room for the breeder both for the foreign and the home-market. The latter will, of course, have to be paid for the trouble, but here co-operation among our tenant-farmers can and ought, in future, to make the burden a reasonable one and the outlay highly remunerative.

CHAPTER VIII

POSSIBILITIES OF THE FUTURE

THERE can be nothing more unsafe than to forecast what is to happen in the future either to the industry which feeds us, or to the agricultural community—the class to which we look for an infusion of healthier blood into that section of the population which has been enervated by the smoke and excitement of large manufacturing towns. Innumerable articles in periodicals and daily newspapers have brought forward every conceivable reason why this country, taught by the experience of the last four years, should maintain a prosperous agriculture in the future, so that the home production of food may be sufficient to protect us against the terrible risk of national hunger which we ran during the war. Judging only by the good intentions, the logic, and the earnestness of the great majority of writers who lay down sound principles of agricultural policy, one would have no doubts as to future prosperity. But, alas, one has to remember that political exigencies have in the past played a prominent part in the direction of public opinion on questions of production; that politicians, while paying lip-service to the principle that food-production is "England's greatest industry," have in practice always supported urban, at the expense of rural enterprise; that the votes of the towns-man must always be more numerous, and therefore more powerful, than those of the countryman. While we may thank God that the heroic efforts of our wonderful seamen have prevented us from receiving in its full bitterness a lesson that would have made us demand fair play for agriculture, we should recognize that if the policy of the past 50 years is to be re-sumed after the war, the efforts of agricultural reformers are useless. Most of the farms where, before the war, intensive production was profitable are not in need of searching reform; there are now, and are always likely to be, some men good enough to carry on wherever circumstances allow of high farming. As to those who may be forced to turn to the robbery of the land as a means of getting a living, experience of the 25 years preceding the war tells us that they know how to steal as well as may be.

One would like, however, to assume that the British public has learnt the folly of relying on the produce of foreign soil for its sustenance while the larger part of its own land is uncultivated or even derelict. I will suppose that my countrymen demand that every acre of these islands shall produce the greatest amount of human food consistent with reasonable economy and that the populace engaged in agriculture shall be enabled to earn sufficient reward for its labour and enterprise to ensure a reasonable number of the best citizens devoting their lives to the industry. Assuming that this happy state of affairs is in prospect, we may consider how cattle breeding (and beef-production) may best be used for the advancement of really productive husbandry in Great Britain.

The first problem to be taken in hand is the systematic sifting-out of all unprofitable stock, so as to secure for the British farmer's service those animals which are most useful in the successful working of his holding. The cattle wanted for export are already well catered for, and the overseas market is a re-munerative one; the prices obtained in the past show that the breeders of this class of stock have been eminently successful, and their work is quite likely to go on concurrently with a grading-up of all the commercial herds wanted for home production. To make a proper division between the useful and the useless, we must have a clear idea of what is actually demanded. The almighty dollar has fixed this point for the export trade. The buyer on this market is very emphatic as to his requirements; if the pedigree, the shape, the colour, the hair and the flesh he demands are not forthcoming, he refuses to buy. In this country we have no such guidance from the English buyer. In previous chapters an attempt has been made to show how this state of affairs has come about, so we may here proceed at once to show how a standard suitable for the home demand could be set up.

The general husbandry of an intensively cultivated country

in which the standard of living is high requires milk, beef, cheese, butter and veal from its horned stock. Having general farming in view, it is certainly economical and convenient to obtain all these products from the same animals. The truth of this is not impaired by the fact that there are always likely to be special types of farms on which an animal is required as being particularly useful for one or other of these purposes only; such cases will probably always be in the minority and, further, as they are more easily investigated, it is easier to set up a standard for them. To fix the standard for the dual-purpose beast, on the other hand, will require our highest efforts and so we will tackle it here. The problem before us is how best to search for the most economical producer of all these commodities among the cattle we now have, how best to eliminate those that are less good than the others, and, while doing so, to disturb as little as possible the smooth running of the agricultural machine.

The yields from the milking stock of this country, as shown by the published records already collected, vary to a much greater extent than is compatible with high production. Milk-recording has "caught on" to a considerable extent, and there seems good ground to hope that, with an increase of State guidance and encouragement in the movement, the farmers will eliminate those cows from their breeding stock which do not yield profitable quantities. The keeping of the records of the cows in pedigree herds is most valuable in creating a supply of bulls from which the owners of commercial cows may find sires to mate with the best milkers in their commercial herds and so make the breeding of good milkers more certain. It would be well to ensure that the influence of deep-milking females should not be lost and that bulls subsidized by any co-operative body should be bred from mothers and grandmothers of proved deep-milking capacity. There are now a few bulls receiving a subsidy, from funds supplied by the Agricultural Development Commissioners, but not only are their numbers small, but also they are not by any means regularly selected with a view to their breeding good milking stock. This is so sometimes even when the members of the association receiving a grant from the Development Fund are the owners of cows used for milk-production. There is room, then, for a great increase in the numbers of these subsidized bulls, and it is of vital importance to insist that, except under special conditions, all subsidized bulls should be bred from dual-purpose stock.

A register of cows giving good yields of milk has wisely been begun by the Board of Agriculture and Fisheries. In this register are recorded, under a number tattooed in the cows' ears, the names of all cows that have given a specified quantity of milk. It is most desirable that this registration should be extended in every possible way. Further, it is most important that the heifer calves from these cows sired by bulls of good milking ancestry should be identified. The reason for this is clear when it is remembered that the calves dropped on the milk-producing farms are often sold on the public market when a few days old. These are frequently bought by dealers who dispatch them to rearers living at very considerable distances from the milk-producing farms. As things are at the present moment, the rearer has no sort of record of the breeding of the calves he receives. He may be rearing the offspring of good or of bad milking stock. If he chances on good ones **he cannot, not knowing the place of their origin**, make sure of securing their fellows at a future sale. Whereas if all calves from registered cows sired by good bulls were marked or branded systematically, they would carry with them the record of their parents' good quality. In this way a demand would be set up for reliable stock of known usefulness, and the breeders would receive some financial reward for producing useful calves. Until this is done, there is no encouragement for the man who cannot rear his own stock to breed good commercial cattle. The system, once started, would not involve very much trouble and it might be made the basis of the levelling-up of every low-grade herd in the country. Almost certainly co-operation between breeder and rearer would follow in the course of a few years and abolish much of the jobbing which is the result of marketing the unhappy little calves, often exposed for days to hunger, cold, and rough treatment while travelling between the place of their birth and the rearer's calf-house. The harm thus done to young calves is one of the greatest difficulties that the calf-rearer has to face, and every means must be sought to lessen it.

Yield of milk, however, is not the only matter that demands attention; there is the question of the cost, in food, of producing the article. Some cows require more food to produce a good yield of milk than others. This point has received considerable attention from workers at the South-Eastern Agricultural College, and at the University College, Reading. They have investigated the systems of management of herds belonging to different owners, and their researches have given results which have been most fruitful in making owners reconsider their methods of feeding. But more than this seems to be required. The capabilities of individual cows seem to promise useful information to the cow-keeper who wants to produce the maximum from the food he gives to his cattle. There is a minimum below which, as we know, it is impossible to go without doing harm. That is to say, if a cow is not supplied with a certain amount of food, according to the quantity of milk she yields, she will either cease to yield well or she will lose flesh to a dangerous extent, or she may do both. But over and above the amount the animal consumes for milk-production, there is a large amount of food used for the maintenance of her body and for the performance of the vital functions going on throughout her daily life. Some cows appear to require more than others for maintenance and for the proper activity of those bodily functions which are responsible for the process by which rough farm foods are eventually converted to the nutritive fluid we know as milk. Cows giving the most milk may possibly be found to be less profitable than those yielding smaller quantities, owing to the consumption of a disproportionately larger amount of food. From my own observations I am inclined to believe that this would be found to be so among our own commercial cattle to a very much greater extent than is at present believed. It is indeed possible that the two things are correlated. For instance, any animal worthy of the name of milch-cow should give an average of 600 to 800 gallons over a period of years, others will give from 800 to 1,000 gallons and so on. It may well be that the most profitable group will not be the one in which the individual animals are the highest yielders. It is quite conceivable that investigation might show that a type of cow yielding a moderately large quantity of milk, when supplied with a moderate amount of food, is more profitable than the very deep milker requiring much more food. On the other hand the reverse may be the truth. In short, there is so much dispute on this point and other matters connected with it amongst those experienced in the management of our English milking herds that it seems certain that systematic and thorough research is required to elucidate the truth.

Beef-production is second only to milk-yield as an essential quality of the beast that is to be an important factor in the high farming of the future; it is little, if any, exaggeration to say that as regards dual-purpose cattle our exact knowledge in this respect is nil. Some years ago I wrote an article on the subject[1] and the trouble involved in collecting a few isolated facts showed how complete was the neglect of systematic work on the subject. It was only by gathering together at great pains the threads of scattered information accidentally recorded that one could get the evidence with which to show the sceptical what was a matter of common knowledge to some, namely the possibility of combining the faculties of beef-production and of high milk-yield in the same animal. As to the profitable combination of the best beef-production and the highest milk-yield in the same animal so as to obtain the highest return for food consumed, we have no reliable evidence about British cattle at all. Yet there can be few matters of greater importance connected with extensive and economical human food-production; for on the vast majority of our farms the fertility which enables bread and other food to be won in great quantity from the land will always be dependent upon a supply of farmyard manure. The cattle, therefore, which will give the greatest return for their own food will be invaluable in keeping up the production of the whole farm by supplying cheaply the basis of all that fertility which leads to increased production. So far, our agricultural com-

[1] "Breeding Milch Cattle and the production of Store-Stock," *The Journal of the R.A.S.E.* 1908, p. 79.

munity has been satisfied with some public exhibitions of very fine fat stock among which were occasionally classes for fat cows. These shows were undoubtedly useful, particularly when the ages and weights of the animals were recorded: unfortunately this was not often, if ever, done as regards the cows. Educational Institutions and Authorities have done much good spade-work in recording weights gained for given rations fed. All such figures, useful as they have been, must be augmented. The foods fed to the stock, the milk-yields of the parents, the milk-yields of the cows themselves, and the returns obtained must all be brought into account if progress is to be made. It may be taken for granted that, for the present, the fat stock show will be held on the old lines as soon as peace conditions permit it, but gradually it will be found possible, I believe, to ensure that the farmers who support these exhibitions will demand more than the meagre information hitherto supplied. It must, moreover, be made possible to get full and exact information about commercial, as contrasted with exhibition, stock, if progress is to be expected. The registration of the cows already begun may, as has been shown above, be extended to a record, by means of branding, of the history of their heifer calves. The systematic marking of their steered calves could be carried on at the same time.

At many of the shows in the North of England and in Scotland there are classes for store cattle. Useful as these are for demonstration purposes, they would be still more valuable if used to prove fully what could be done by breeding being confined to the most useful type of animal. If it was known how the store-stock at these shows, as well as those at some public sales, were **bred** and **fed**, their **ages** and their **weights**, we should be able to get much valuable information at present unobtainable. Such general information would be a reliable check on results obtained with small numbers of animals kept under trained observation during feeding trials. The returns obtained as produce must be the criterion of successful breeding and feeding operations, and once the practising agriculturists fully realized that certain brands of stock gave them better results than others, they would buy them; thus better prices would reward those who succeeded in breeding the profitable calves, and the food-supply of our country would be increased. The present lack of systematic method is very far from encouraging the cow-keeper who sends better calves to the market than his neighbour; for, generally, they pass through several hands in the course of their lives, and all trace of their origin is lost. A farmer getting hold of a really useful drove or "bunch" of young cattle seldom knows, and has no means of discovering, their origin, and so cannot, with the best will in the world, go and pay a good price to ensure a supply of the same class of cattle. Extension of pedigree recording and systematic research work *together with* co-operation would put an end to such a state of affairs.

Cow-beef is an important item, even in the dietary of the British public, though to a lesser extent than on the continent. Yet no systematic inquiry has been made into the best method of converting the lean milking-machine into an edible carcase of meat. Some people milk till nearly all the flesh is off the bones and then turn their cows out as "stores" for a long period of feeding or fattening. Others feed very high rations and so induce fleshiness during the period of lactation, so that the cow is fit for the butchers as soon as she is dry. A third system is a compromise between these two methods, feeding for beef being begun about half way through the milking period and ending a short time after the cow has stopped yielding. Granting that there always will be *circumstances*, as undoubtedly is now the case, to decide which of these three methods is best on a particular farm, it must be useful to ascertain which method produces the most meat for a given outlay in feeding-stuffs. Furthermore, we cannot decide which animal is the most profitable until we know how much influence the previous deep-milking record has upon meat-production once lactation is over. On these and many other points knowledge is, to say the least, very limited. Accumulation of exact data alone can show us whether there is not room for very great improvement in our present methods of production through our milch-cows as well as our other cattle.

This question of beef requires much deeper investigation than a mere record of the live weight of the stock produced.

Different animals yield amounts of meat that differ very greatly in proportion to their live weight. Under our present system of marketing this point concerns the consumer more directly than the farmer. One butcher's beast bred and treated in exactly the same way as its neighbour may yield 54 per cent. of carcase, while the other will yield 58 per cent. Under our present system the farmer, as a rule, now gets the same price for both animals, for they appear to be similar; but it cannot be profitable to the country to go on paying the same price for the beast of lower yield. We want to find out, also, what is the most economic stage of the animal's life, or what is the most profitable condition at which to slaughter. An inquiry[1] which Dr F. H. A. Marshall and I carried out in 1917 shows clearly that not only do animals differ in a very marked degree from one another as regards the proportion of meat-yield to total live weight, but also that the profitableness of the carcase, as a means of supplying human food, is a very varying factor. The inquiry further showed, beyond any possible doubt, that even among the experienced, knowledge was nebulous and belief constantly erroneous about this very important matter. The subject is a difficult one about which to acquire accurate know-ledge, and consequently will require skilled attention, time, and money to secure clear guidance for consumers and stock-breeders. In the past so little attention has been given to the subject (perhaps on account of the difficulties involved) that it may almost be said that there lies an absolutely new field for research ready to be worked by the expert in animal husbandry who wishes to select the best from among the cattle of this country.

Investigation along these lines of meat and milk-production might not only prove of great value in enabling us to choose the most profitable animals for stocking and breeding, but should also be very useful as a means of imparting better technical education. In fact it may be said that unless the

---

[1] Since the above was written a summary of the report on this inquiry was published in the *Journal of the Board of Agriculture* for September 1918; it was therein announced that the full report would appear in due course; subsequently, however, it was given out that the publication of the full report was postponed.

farmers, more especially the younger men, are made familiar with the work during the whole of its progress, the results may well be abortive. One of the greatest triumphs that systematic investigation has achieved in husbandry has been in connexion with the increased production from the land obtained of late years in Denmark. In these investigations the problems of pig-meat production played a very important part. The same eagerness for knowledge and the same educational effort must be shown in Great Britain and Ireland if we are to produce enough from our land to make us feel that unnecessary waste of potential food is being avoided, or, at any rate, not going on in the same wholesale manner as immediately before the war. Many trials would be needed from which to obtain accurate and informing results. If they were widespread, they would have an inestimable educational value. If they were brought to the very doors of our farmers, they would not only eventually lead to much useful information being acquired, but also to an improved intellectual outlook among the agriculturists who were familiarized with the work. No one who has worked long among the farming population can fail to observe that the outlook of many practitioners is a narrow one, and every effort must be made to widen it, for it is on them that we depend to win food from the land. Experience shows more and more clearly that to tell the average English farmer about the methods adopted in foreign lands is useless. He is by nature too practical not to realize that different conditions of farming demand different treatment; and his education and training in systematic work are still too scanty to enable him to borrow ideas from the foreigner, and adapt them to a practical usefulness in his own farming. We must always bear in mind that while a small minority of our agriculturists are as good as any farmers are ever likely to be, the great majority of our men are not accustomed to *thinking* of increased production; for they have been forced to fix their minds upon the one problem of how to get a profit without any regard to the usefulness of their goods to the nation as a whole. The older men may be too firmly set in this one idea to change, but it is not so with the younger ones. Experience of nearly 20 years shows that, greatly as a narrow

upbringing may handicap boys and young men bred on the
land, there is among all classes of them splendid material
for the educationalist's work; and, knowing this, I personally
have a very firm belief in the possibilities of using research
stations all over the country and proving their value both as
a means of training our men and of securing for us the most
valuable cattle.

It must be freely admitted that in the past we have had very
little solid fact on which to base our teaching. We have too
easily accepted conjectural estimates as truth, and from such
conjectures divisions of opinion naturally arise. It must indeed
be difficult for the student to form a judgment when the opinions
of authorities differ. For instance, it is often maintained that
certain outward indications are evidence of good milking
qualities in the cow. One of these indications concerns the
shape of the chine, that part of the body where the tops of the
two shoulder-blades converge towards the points of the spine.
A good beef-animal is required to be wide and level at this
point, and the bones of the shoulder-blades should run as far
as possible parallel to one another. It is said to be an indication
of milk-producing capacity for a cow to be rather narrow, inclined to be sharp, and for the shoulders to converge on one
another so as to make the whole chine wedge-shaped. A very
well-known authority does not believe in this theory. He has
pointed to the Aberdeen Angus breed'as a proof of his contention,
saying that this animal though well known as belonging to a
beef-breed and consequently wide at the chine is nevertheless
often a good milker. On the other hand, I personally incline
to believe that a fine chine does indicate a capacity for milk-
giving. I have told many students that I look upon the Aberdeen
Angus as a breed **which supports my view,** for I tell them it
has always appeared to me that the Aberdeen Angus cow that
proved the exceptional milker among her tribe was exceptionally
fine at the chine! This instance is but one example of many
such contradictions that will constantly face the student while the
present indefiniteness of knowledge prevails. It will be very poor
encouragement to the agriculturist who is anxious to increase
the productivity of his cattle if no real effort is made to obtain

for him definite information about all kinds of possible improvement. He cannot, indeed, do his part thoroughly unless a widespread effort is made to help him by research into the many uncertainties of the subject.

There cannot, it appears to me, be a more splendid field of activity for the landed proprietor of this country than to use his wealth and influence for the support of this work. Surely nothing but improvement would result to the home-farms of England if their owners would use them for the advancement of knowledge and education by means of research. A national scheme of research, combined with an educational movement, would be both better in itself and much less costly, if all possible help were given to it by the landlords. To the individual, moreover, it would not be a very expensive means of helping the nation; for the chief requirement would be a man on each farm willing and competent to make the weighings, superintend the feeding of rations, and keep the records—and almost any farm-manager could carry out what is wanted. Is it then too much to ask that the proprietor should take sufficient interest in the production of food from the land to see that such help is carefully given? Associations of agriculturists do this on the continent, and if the landlords took a real lead in agricultural matters they would soon see that their example led to the tenants doing still more, and thereby helping the country and themselves. Some land-owners have helped in this way, and the good they have done shows what great benefits would result if the practice became general. It would be the duty, of course, of the central Research and Teaching Institutions to prepare the groundwork of all methods of investigation, to keep a general supervision over the work, and to digest and correlate the results; in this way the detailed work would not be too great a tax on willing helpers.

There are many other problems connected with the cattle industry and its productivity which require more knowledge based on research. We shall omit those concerned with pathology, for, important though they are, veterinary matters are outside the scope of this volume; but akin to them are many problems of physiology about which the stock-breeder is now very much in the dark. For instance, at the present moment no one can say whether the powers of reproduction are hindered, or not, by pushing the lactation functions too far. The irregularities of breeding are constantly interfering in so many ways with the profitableness of our live-stock, that an effort is urgently needed to allow, and even to encourage, the man of science to ascertain whether he can to a certain extent help the farmer in the control of breeding. It surely ought not to be, as it is now, a constant source of dispute as to what is the right age to breed from a heifer. There are few districts in which practice does not greatly differ in this respect, even when the animals are all wanted for exactly the same purpose. If there is no benefit to be derived from the practice, it is surely a bad thing that 50 per cent. of our female cattle are allowed, without any adequate reason, to remain useless as milk-producers for six months longer than the other half; yet no one familiar with prevailing methods can deny that this is the case. It is of the utmost importance that the regular supply of milk should be kept up in the kingdom; yet the factor of calf-production, which regulates the supply, is so uncertain that cow-keepers often suffer loss through being obliged to keep a surplus of breeding animals to ensure that the supply of milk does not fall below a certain minimum. There is a long list of questions concerning fertility that require investigation. Recent research into the question of fecundity in sheep[1] leads to the hope and belief that research work dealing with such matters might lead to results of great benefit to the stockman.

There remains another question, namely, whether we, as breeders, have yet obtained the animal that most nearly approaches perfection in meeting our requirements. Mendel introduced entirely new conceptions of the possibilities of breeding. The science of genetics, as the study of the laws of inheritance is now called, might be utilized to combine more good qualities in the same body. This has been done by Biffen and others as regards farm-crops; why, then, should we not hope for the

[1] Marshall, "Fertility in Scottish Sheep," *Transactions Highland and Agricultural Society of Scotland,* Vol. xx, 1908.

same useful work in connexion with living animals? But there is much to be done in preparation for this final effort at improvement. We have no very clear idea of why one animal is good, and another bad, for a certain purpose. By selection of those animals which seemed good for their requirements the breeders of the past have made a very considerable advance towards perfection. If we made it certain that all our husbandmen produced only the best of what we already possess we should have obtained a very great deal; it may be that this would be the best standard to aim at in the immediate future. But, on the other hand, there are so many imperfections in the dual-purpose animal of the present day that it would seem folly to neglect such future possibilities as might accrue from investigating the wonderful laws which Mendel first brought to the notice of science. For instance, it is generally held that the carcase of meat obtained from an animal belonging to one of the pure-beef breeds is more useful to the butcher than that obtained from an animal of a variety that has been selected through many generations for deep-milking qualities; and this notwithstanding that the breeders of deep-milking cattle had sought by all means in their power to breed with a view to good fleshing qualities as well. The new science of genetics might help us to combine the best beef-production with the most valuable yield of milk in the same animal; but before this can be done it would appear to be essential to know what are the characteristics of a good carcase and what factors determine them. For some years I have sought in vain for information on this point. Empirical opinion seems to guide the trade into certain channels along which a selection is made, but nothing definite enough has been established to enable a research-worker even to start a definite line of improvement. As to the part the skeletal structure, the quality of bone, the development of certain muscles, the quality of muscle fibre, the proportion, quality and manner of depositing fat, all play in providing a good or bad carcase there is no information on which to work. Apparently the very essence of the whole subject, the question of what is good and what is not so good, has been simply a matter of opinion, or even fancy, on the part of the individual, or of a group of meat-salesmen or butchers. There are many points, some of them perhaps very subtle, which must be worked out before it is possible for the man of science systematically to superimpose one good quality upon another.

In this connexion there is a further subject to be studied: namely, to determine which features of the living animal denote the characters we require. The plant-breeder can retain a large number of seeds and easily study their properties through the medium of the plants they produce after a short period of growth. With calves the matter is very much more difficult and the expense almost prohibitive. Of a hundred bull-calves dropped no more than 5 per cent. are usually retained for breeding purposes. To keep the whole hundred and investigate their qualities as progenitors would be altogether too cumbersome and would require too great an outlay to allow of much hope of return. But if the touch, the eye, the weighbridge, or measurement, or all these combined could tell us definitely a few days after birth that a calf was the right sort to be kept as a sire, systematic breeding would be brought within the range of possibility. Our present knowledge, great as it undoubtedly is, is not sufficient to form a basis for the accurate work which science demands.

Many such problems could be quoted, for every one who has given serious thought to the subject has some particular point in mind as a subject of research which should lead to improvement in the economic value even of animals now thought to be good. The great object to be attained, however, is the elimination of all unprofitable animals, and the collection and use of evidence to demonstrate the folly of breeding and keeping a large proportion of the animals we now unfortunately possess. Education must in the future carry conviction that the fully-developed cow, yielding no more than 500 gallons of milk per annum and at the same time being but an inferior animal for the shambles, is unworthy of the husbandry of a civilized nation. A great effort will be needed to teach a large proportion of our farmers, long engaged in rearing cattle of this class, that their only source of profit (and that unsatisfactory) has been robbery from the land. The stock-breeding industry deserves a good reward.

If, however, the British public is prepared to pay, it is entitled to "call the tune"; and the key-note to be struck is that no profit can justly be allowed to be made from any class of stock that does not adequately and economically help to fill the national larder.

CHAPTER IX

PHYSIOLOGICAL[1]

THE value of meat as an article of diet depends upon its containing concentrated nourishment in a form which can be easily dealt with by the organs of digestion, and so as to yield only a minimal amount of faecal residue. With meat-eaters the nutritive processes consist in a conversion of the flesh and blood of herbivorous animals into the flesh and blood of those consuming them, the constituents of the tissues being approximately the same in the two classes of animals. An elementary study of the comparative physiology of Carnivora and Herbivora is sufficient to convince one of the general truth of these statements.

In carnivorous animals, such as the dog and cat, the digestion and absorption of the foodstuffs are practically confined to the stomach and small intestine. The caecum or "blind-gut" is reduced to a mere vestige, and the colon (which together with the caecum and rectum constitutes the large intestine of mammals) is small and short and without any of the complications found in the horse and other Herbivora. In man the caecum is likewise vestigial, consisting of little more than the appendix vermiformis, so often removed surgically and without any subsequent detriment to the powers of digestion.

On the other hand in the Herbivora the digestive apparatus is enormously developed. In the horse the caecum is so immense as to occupy the greater part of the abdominal cavity, while its length may be greater than that of the body. At the same time the colon is a vast structure, five or six times as large as the stomach, and double throughout part of its length. In the rabbit and many other rodents also the large intestine is of enormous dimensions relative to the size of the animal, and the same may be said of the herbivorous marsupials.

In the ruminants, of which the ox, the sheep and goat are examples, the large intestine relative to the size of the animals

---

[1] By F. H. A. Marshall, Sc.D.

is considerably bigger than in the Carnivora, though smaller than in the horse. The enormously distended rumen or paunch and the other alimentary compartments at the fore-end of the gut, compensate in ruminants for the somewhat reduced large intestine. Thus it may be said that the rumen in the ox is the functional though not the anatomical equivalent of the caecum and colon in the horse or rabbit.

An examination of the faecal or undigested residue evacuated through the anus tells the same story. Bischoff and Voit[1] pointed out long ago that the faeces of the Carnivora when fed on lean meat consist almost entirely of excretory products which are got rid of through the wall of the gut, and are strictly comparable to the organic matter of the urine. Only traces of undigested food occur. "Metabolic products" consisting of altered residues of digestive fluids, mucus, etc., are evacuated in small quantities even in a condition of abstinence when the gut is empty of food —the so-called "fasting faeces," and according to Armsby[2] "the consumption of highly digestible food—e.g. lean meat—does not seem materially to increase" even the fasting faeces, "though when food containing indigestible matter is eaten it is believed they increase in quantity." Moreover, unless the meat is given too freely the faeces of the dog never contain undigested muscle fibres. Dogs when fed on a diet of meat defaecate (i.e. empty their bowels) at intervals of from two to four days, and as already remarked the amount of residue ejected is extremely small, perhaps 3 or 4 oz. Meade Smith[3] has calculated that a dog weighing 35 kilogrammes and fed on a half to two and a half kilos of meat, evacuates from 27 to 40 grammes of faeces containing only 9 to 21 grammes of solids. "Therefore, it may be said that with a flesh diet, only one per cent. of the amount of solids taken with the food escapes from the body in the form of faeces." On the other hand in herbivorous animals defaecation takes place at very frequent intervals, and large quantities are passed *per anum*. Thus the horse evacuates every three or four hours,

[1] Bischoff and Voit, *Die Gesetze der Ernährung des Fleisch Gressers*, 1860.

[2] Armsby, *Principles of Animal Nutrition*, New York, 1906.

[3] Meade Smith, *The Physiology of the Domestic Animals*, Philadelphia, 1889.

and the daily average quantity of faeces is 30 lb., but may be double as much. With cattle the amount is still greater, and averages 75 lb. The sheep and pig discharge about 4 lb. of faeces daily.

"Animals living on a mixed animal and vegetable diet, like man, pass daily about 130 grammes of faeces, containing 34 grammes of solids, which will represent about 5 per cent. of the solids taken as food. When the vegetable diet is in excess, this may rise to 13 per cent., so that only seven-eighths of the solids are finally absorbed.... In a man fed on a milk and meat diet only 2½ to 10 per cent. escapes in the faeces; with a vegetable diet, rice, bread and potatoes, the loss may amount to 30 per cent., though the carbohydrates are almost completely absorbed, only 1 per cent. being lost, and only 5 per cent. of fats escapes absorption."

An examination of the urine of different classes of animals is in certain respects equally suggestive. Thus in man, indican though often present is not a normal constituent of the urine, but is regarded as evidence of "indigestion" or incomplete metabolism. But in the horse or ox indican is invariably to be found in the urine and often in considerable quantity.

It is, of course, incontestable that the nutritive habits of animals can be changed under artificial or unusual conditions, that many Carnivora can be got to subsist on a purely vegetable diet, and that herbivorous animals may be induced to feed upon fish or even upon flesh, food-substances for which their dentition is entirely unsuited. Moreover, by cooking the food and thus rendering it softer and reducing it to a condition in which the digestive fluids can play upon it more freely, and such that the undigested residue can be more easily passed onwards by the peristaltic movements of the gut wall, the disadvantages of an abnormal diet may be reduced. Nevertheless, when animals fed in an unusual way are given the chance of returning to their natural food they almost invariably do so.

This is not the place, even if space permitted, for a general treatment of the fundamental principles on which a rational system of feeding must be based, but enough has been said to show that the consideration of a dietary is not merely a matter of calories; there are other factors of importance, both physiological and psychological, which must be taken into account; even the palatability of a food is not a matter to be dismissed lightly, for palatability is an assistance to appetite, and the existence of appetite is functionally correlated with the secretion of some at least of the digestive ferments. Thus the experiments of the distinguished Russian physiologist Pavlov[1] showed that the sight, smell, and taste of food, acting reflexly through the nervous system, are the exciting cause of the secretion of the saliva and the gastric juices, and that the glands producing these digestive fluids are brought into functional activity under the stimulating influence of appetite before the food is swallowed or even before it enters the mouth.

Now of the food substances which enter into the diet of a normal man in a temperate climate, meat is one of the most important. This may be partly the result of habit, for meat is not an irreplaceable necessity (at any rate for many individuals) as successful vegetarianism has proved. But although it is probably true that in pre-war days the amount of meat consumed by certain classes of society was wastefully excessive, it is indisputable that for the normal man a moderate allowance eaten daily is an unqualified advantage, since its effect is to decrease the labour of digestion and to reduce the faecal residue. The desirability therefore of ensuring a sufficient supply of home-grown meat is a matter of the greatest national importance from the point of view of preserving the strength of the people and maintaining in times of crisis and difficulty their morale in full measure.

Meat, in the purely restricted sense of the term, consists of muscular tissue, together with other tissues which are closely associated with it, that is to say the connective tissues which bind the muscle fibres together, the slight amount of fat which is deposited in the connective tissue, the blood vessels which supply nourishment and carry away the waste products, and the minute nerves which connect the muscles with the brain or spinal cord, and provide a co-ordinating mechanism for the

[1] Pavlov, *The Work of the Digestive Glands*, translation by W. H. Thompson, London, 1902.

movements of the living body. The muscular tissue itself, which constitutes by far the greater part of the substance of the meat, consists of striated fibres which taper at both ends. Each of these fibres is a single cell provided like other cells with a nucleus and enclosed by a membrane forming the wall of the muscle fibre. The muscle fibres running parallel to one another are united into bundles by connective tissue, and the bundles themselves are further bound together to form the entire muscle. The meat derives its red colour from the haemoglobin or red pigment of the blood.

Butchers' meat in the more general sense consists of the prepared carcases of oxen, sheep, and pigs, and of the young of these animals, but in the present work we are restricted to the consideration of beef-production. In the living animal, two phenomena contribute to this object, growth and fattening. In a general popular sense these terms are understood, but they require exact definition or at any rate description. Both processes involve the addition of new tissue, that is of living matter or protoplasm, but whereas the term growth is usually confined to that increase which is correlated with the development of the individual from birth to maturity, fattening is a process which may be reversed many times over at any period throughout life, a fat animal becoming lean and a lean animal fat. Strictly speaking the term fattening ought to be restricted to the addition of adipose tissue, but it is often used much more loosely so as to include the putting on of more muscle when that process is accompanied by an increasing accumulation of fat. In a similar way the term growth is sometimes employed to express the development of muscular tissue (that is of meat) even late in life when skeletal and developmental growth has ceased. Such addition of muscular tissue, as is well known, is the usual result of increased use or exercise which generally implies also an increase in the food consumed.

In cattle the skeleton or bony frame gradually goes on growing until a certain age, usually about five years, after which time it remains the same. In oxen, as a result of castration, the limb bones go on growing for a somewhat longer period, since one of the effects of the removal of the generative organs is to prevent the end parts or epiphyses of the long bones from ossifying so rapidly; the fact that these retain their cartilaginous character for a longer period admits of the shafts of the bones continuing to grow until prevented from so doing by hard, dense, ossified structures at their ends.

Growth and fattening in actual practice proceed simultaneously, but at a certain stage developmental growth and increase of protein tissues cease (excepting in cases where further development of muscle is due to increased exercise) and any subsequent addition is due exclusively to increase of fat. Such a condition is sometimes spoken of as the "par" condition[1]. An animal in "par" condition is in a state of nitrogen equilibrium, the amount of waste nitrogenous material eliminated through the kidneys being equal to the amount absorbed in the digestive tract.

What it is that determines the precise stage at which the par condition is arrived at is still an open question. It can only be said here that it is apparently an hereditary characteristic varying in different individuals and still more in different varieties and species. All animals follow the same general laws of growth. In the earliest stages development is very rapid, next there is an enormous fall in the percentage of growth, then the decline becomes gradual (varied only in entire animals by fluctuation at about the period at which sexual maturity is acquired), till finally developmental growth ceases altogether. The laws of growth are believed by some to be related to cellular changes, the nuclei becoming relatively smaller and gradually losing their rejuvenating capacity. Hence their power of forming new tissue is gradually reduced. For as stated already the deposition of fat in the cells is not a case of true growth, but may partake rather of degenerative change.

Fat is formed mainly in the connective tissue cells, that is, in those tissues which bind together the various organs of the body and hold them in position, filling up interstices and forming a net-work which tends to cover or pervade the more highly differentiated tissues. The fat is produced by a process of absorption from the blood vessels, the protoplasmic contents

[1] Murray, "Meat Production," *Science Progress*, April 1918.

of the cells becoming gradually displaced while the nucleus is pushed to one side, coming to lie against the cell membrane. The fat cell formed in this way grows to be very much bigger than the original protoplasmic cell, and the large collections of fat cells become packed closely together, being only separated by very thin membranes and forming adipose tissue which may be many cells in depth. This is what occurs in the normal process of fattening. In other cases where the fat formation has been carried to excess the cells of other kinds of tissue such as muscle become the repositories of fat which interfere with the discharge of the normal functional activities. Such is the condition occurring in fatty degeneration which when it affects the tissues forming the wall of the heart is a source of danger to the life of the animal.

Apart from definite degenerative change which does not usually occur until comparatively late in life, the adipose tissue formed in a fattening bullock may be considered under three heads.

Firstly, there is the fat laid on in and beneath the dermis or deeper layer of skin and upon the underlying muscles. Thus if we take a rib of beef as cut in a butcher's shop we find that the two principal outside muscles which are seen in cross-section in the "joint," the trapezius which lies dorsally (i.e. towards the spine of the vertebra) and the latissimus dorsi which lies ventrally beginning where the trapezius ends, are covered by layers of fat which vary in thickness, and are a rough index of the condition of the animal. We note also that the fat covering the trapezius muscle is invariably thinner than that on the outside of the latissimus dorsi, and that in lean stores the trapezius, especially in its upper or dorsal part, is hardly covered at all. We note also that the other muscles (e.g. the ilio-spinalis or the muscle forming what butchers sometimes call the "eye" of the joint), which go to make up the meat, are separated from one another by fascia and connective tissue together with fat which like the superficial fat varies in amount according to the beast's condition. This fat, together with that referred to above as covering the outside of the joint, we speak of as the "gross fat."

Secondly, we generally see a certain amount of fatty tissue

within the muscles and lying in the connective tissue between the bundles of fibres in streaks or irregularly arranged patches. In lean beasts, especially those which are immature, this fat is so small in quantity that it cannot be discerned with the naked eye at all. When developed to any perceptible extent, it gives rise to what is commonly called the "marbling" of the meat. Marbling is a quality which adds greatly to the value of beef, especially in the older animals where the muscle fibres are tougher and less succulent, and it renders the meat easier to cook and easier to carve. Meat, therefore, which is well "marbled" is superior in "quality" and more attractive than lean meat which is not permeated by fat in this way, but in young animals marbling is of less importance than in a six-tooth or eight-tooth bullock on account of the greater tenderness of the meat.

Lastly, there is invariably a greater or less amount of fat accumulated within or immediately surrounding the body cavity. This consists of the so-called kidney fat, and the fat which lines the peritoneum or covering membrane of the wall of the abdominal cavity, and the "gut" fat which is deposited in the omentum and mesenteries or membranes which support and enclose the alimentary canal and hold it in position. The so-called "apron" or "caul" consists mainly of mesenteric fat which may be very excessive.

The respective degrees of development of these three classes of fat in beasts of different breeds, types, ages and conditions have formed the subject of a recent investigation by Mr Mackenzie and myself[1]. That there is some correlation between them is only what might be expected, but the extraordinarily wide range of variation in the proportion of "gross" fat to marbled fat and in the connexion between either of these and the yield percentage (carcase to liveweight) is far more striking than the correlation which is only seen at all when the average of a large number of beasts is taken. There are, however, certain outstanding facts which may be briefly recorded here together with the conclusions that must be drawn from them.

[1] Mackenzie and Marshall, "Beef Production" (Abstract), *Jl of Board of Agric.*, Vol. 25, 1918.

It had been hitherto assumed that there is a direct relation between the proportion of fat found in an animal and its percentage yield (i.e. the percentage of carcase weight to liveweight). Thus the late Sir John B. Lawes[1] constructed a table in which the beasts are divided into five groups, as follows:

(1)	Stores	50–51 %
(2)	Fresh stores	52–53 %
(3)	Moderately fat	54–57 %
(4)	Fat	58–62 %
(5)	Very fat	63–65 %

The results of the investigation above referred to show that this correspondence is very inexact. For example, beasts with a percentage yield of 56 may show a fat proportion (as determined from an analysis of the 7th rib) of as low as 15 per cent. or as high as 33 per cent. Again, beasts with a fat proportion (in the 7th rib) of 20 may have a yield percentage of anything from 53 to 57. Furthermore, it was found that whereas a beast with a fat proportion of 28 had a yield percentage of 62, the cattle showing a higher proportion of fat had lower yield percentages, those showing the greatest amount of fat of all, namely 33 per cent., having in each case a yield percentage of only 56. This is because the offal and other internal fat which does not enter into the carcase weight necessarily reduces the yield percentages. These and other facts, which are too numerous to set out here, show that any estimate made upon the living animal and based on a feeling of fatness is a very unreliable way of arriving at the actual percentage yield. Judgment depending on such a method is necessarily faulty, for the person whose hand touches the beast is apt to set far too high a value upon it if he feels fat to have been laid on in any quantity.

The amount of intermuscular or marbled fat is only very roughly correlated with the amount of gross fat, that is to say, some individuals as their condition improves tend more especially towards an addition of gross fat, and some show a more marked increase of marbling. These differences do not neces-

[1] Lawes, "Tables for estimating Deadweight and Value of Cattle from Liveweight," published for the author by the Royal Agricultural Society of England.

sarily depend either upon breed, or age, or food, for some beasts which have been bred in the same way and treated identically vary considerably in their respective proportions of gross and marbled fat. Apart, however, from this individual variation, analyses have shown that age and breed are definite factors in the correlation between the fat added on the outside of the muscles (the gross fat) and that put on amid the bundles of muscle fibres (the marbled fat).

Thus with Shorthorn beeflings (with no broad teeth) the greatest amount of marbled fat (as found in the ilio-spinalis muscle) was 18 per cent. in the 7th rib and 13·4 per cent. in the last or 13th rib (i.e. the wing rib) as compared with 33 per cent. of gross fat in the 7th rib. All these amounts were exceptionally large and were the result of specially good feeding. Shorthorn bullocks with six or eight broad teeth upon the other hand showed the following analyses:

Yield Percentage.

Gross Fat	No. of broad Teeth	Marbled fat in 7th rib	Marbled fat in 13th rib
26·8	6	30·8	22·7
27·3	6	26·3	15·7
27·2	6	32·2	22·0
30·9	8	27·8	20·9
33·7	8	23·3	17·2

The analyses above recorded point to another conclusion of interest, namely, that in older bullocks though the gross fat may increase the amount of marbling does not increase; as a matter of fact in the two eight-tooth bullocks (the only two analysed) the average amount of marbled fat was considerably below that found in the six-tooth bullocks, although the gross fat had increased. In view of the importance attached to marbling by the butcher this observation is noteworthy.

The conclusion that after a certain stage of fattening marbled fat does not increase so readily as gross fat has been arrived at in another way. As a result of an analysis of the 7th and 13th ribs in some 90 beasts of different ages, breeds, types, percentages and conditions, and after dividing them into four groups according to their yield percentage, the following results are reached:

	Under 53 %	53–54 %	55–56 %	57 % or over
Gross fat	15·5	18·4	21·5	24·4
7th rib marbling	11·8	12·9	17·3	20·4
13th rib marbling	6·4	8·6	12·5	13·9

It is thus seen that in both ribs the marbled fat in the second group (53–54 per cent.) does not greatly exceed that in the first group, but that in passing to the third group (55–56 per cent.) there is a pronounced rise of fat. In proceeding to the fourth group in which the yield percentage reached 62 and was often 59 or 60, the rise in the marbled fat in both ribs is a good deal less than in passing from the second to the third group. It must, however, again be emphasized that these figures only represent the averages, the individual variation being considerable.

There is further evidence that the degree of marbling is a racial characteristic, as the following analyses from four Hereford bullocks of approximately the same age show:

Gross fat	No. of broad teeth	Marbled fat in 7th rib	Marbled fat in 13th rib
20·8	4	19·1	14·8
23·6	3	16·2	9·5
24·9	4	17·9	12·7
31·3	4	18·3	13·5

These may be compared with the analyses of two Lincoln Red bullocks:

Gross fat	No. of broad teeth	Marbled fat in 7th rib	Marbled fat in 13th rib
27·5	4	27·6	20·8
31·9	4	34·6	20·8

Speaking generally, however, the number of beasts so far analysed is hardly great enough to admit of final conclusions being arrived at regarding marbling as a characteristic of breeds. It is evident, however, that the variability both racial and individual which recent investigation has disclosed opens up wide possibilities of future improvement by the ordinary methods of selective breeding. It should not be difficult to discover typical strains of cattle in which the capacity of the flesh to marble is above the average, and once this has been done it should be only a matter of a few generations to obtain something so closely approaching fixity as to be of permanent practical value. Such work could best be carried out at a joint Institute for Animal Breeding and Animal Nutrition, where this and many kindred problems could be effectively studied, and we know of no way by which anyone who has the will and the means to benefit our national live-stock industry could do so with a better assurance of success than by providing an endowment for such a scheme.

Before we leave the subject of marbling there is another point of both physiological and practical interest which is worthy of our attention. The analyses of the ilio-spinalis muscle in something over 80 beasts have consistently shown that the amount of intermuscular fat diminishes in passing backwards from the anterior ribs to the posterior ones, and that the proportion present in the wing rib and first cut of loin is invariably considerably smaller than in the middle rib. In view of the importance attached to marbled meat this result might seem surprising, since the loin and prime ribs are universally regarded as the best joints in the carcase. But it has been pointed out above that the fat occurring in marbled meat is not deposited within the fibres but in the connective tissue which surrounds and binds together the bundles. Moreover, the meat which is near or in immediate anatomical connexion with skeletal structures contains more connective tissue than that which is further away. This connective tissue when well developed is *one* of the factors (but only *one* of them) in making meat tough. Here at once we have the explanation as to why the joints cut from the posterior ribs and loin are apt to be more tender, and at the same time less marbled than the joints taken from the anterior ribs or elsewhere, since the muscles composing the meat of the hind ribs and loin for the most part have their insertions and attachments at the fore end and hind end of the animal. But it must be clearly understood that although marbling appears to depend upon the existence of connective tissue, the converse of this does not follow, for, as we have seen above, the transformation of fibrous or elastic into adipose tissue is dependent upon "condition," besides being probably a racial or at any rate an individual characteristic. Apart from the degree of development of connective tissue, "toughness" and "tenderness" in meat are due largely, as is well known, to the denseness of the muscle fibres and the thickness of their cell walls, characteristics which are developed by use, that is, by muscular exercise.

The muscular development necessary for remunerative beef-production must depend further upon the shape of the skeletal framework, that is to say, it is an anatomical characteristic of particular breeds (see Chap. X). Moreover, beef and milk-production may be usefully combined in the same breed or strain, and of this fact the living proof exists in the "dual-purpose" Shorthorn and Lincoln Red cattle, and to a less extent in certain other breeds. But to what degree a good beef carcase can be united in the same race with a deep-milking propensity is still an open question. It cannot be definitely said that the two characteristics are physiologically incompatible, though it is probably quite correct that for reasons of nutrition they cannot easily co-exist in the same individual. On the alternation between the growth of dermal fat and the secretion of milk we find Sheldon writing as follows: "On animals that are well adapted for both milk and beef, there will always be soft, velvety skin, which will feel mellow to the touch as if it rested on a second underskin like a cushion. This 'underskin' consists of a network of cells, called 'cellular tissue,' and when a cow is not in milk fat soon accumulates in it, and forms the 'quality' or 'handling' which indicates the extent to which she may be considered fit for the butcher. If the cow is in milk this fatty accumulation in the cellular tissue goes instead to form cream in the milk[1]." These statements are crudely and inaccurately expressed, since there are reasons for believing that the fat contained in milk is manufactured in the cells of the mammary gland, and not carried thence from elsewhere, but nevertheless the general contention implied (namely, that the increased nutriment required for an organ in a state of functional activity involves a depletion of tissue elsewhere) is not without a solid substratum of truth.

[1] Sheldon, *Dairy Farming*, London.

A study of the correlated changes which characterize the oestrous or sexual cycle in female animals supplies evidence of the truth of the same principle. It has been observed that in the early stages of pregnancy when the foetus is still small and the developing mammary tissue inconsiderable, there is often a tendency to put on fat. This tendency is associated with the cessation of the "heat" periods, a cessation which implies the arrest of ovulation (i.e. the discharge of eggs from the ovary) and those other characteristic phenomena of oestrus which manifest themselves externally in the display of sexual feeling, and involve a considerable expenditure of vital energy. Subsequently, however, when the need for nourishing the rapidly growing calf involves a serious drain upon the mother, and when the developing mammary tissue is in a state of great activity preparatory to parturition, the tendency to fatten induced by the "settled" condition of early pregnancy ceases, for the functional activities of the mother as pregnancy proceeds tend more and more to become subservient to the needs of the offspring.

The changes which occur periodically in the ovary are clearly intended (to speak teleologically) to secure this end, that is to say, in the normal condition they are regulated with a view to fecundity, and to co-ordinating the activities of the generative organs and mammary glands so as to produce a minimal amount of waste in the energy expended upon these processes.

The two ovaries (which are placed symmetrically on the right and left sides of the body cavity) are the essential reproductive organs of the female, just as the testicles are the essential male organs. The other female organs (the oviducts, the uterus, womb or "bed," the vagina or common uro-genital passage, etc., and the mammary glands) are to be regarded as accessory, and the periodic changes which take place in them are dependent upon the varying activities of the ovaries. The ovary is like the mainspring of a watch, it governs all the sexual functions and controls the oestrous rhythm, and if it be removed the cycle ceases, and sooner or later atrophy sets in among the accessory organs. Just as the rhythmic movements of the hands of a watch or clock, both the short and long hands and the second hand, and, if there be one, the partially detached mechanism of the striking apparatus, are dependent upon the mainspring, so the sexual cycle, both the complete annual cycle and recurrence of the three-weekly periods, and also, though in a more remote and complex way, the recurrent changes which take place in the mammary glands and the growth of the uterus in pregnancy, are controlled and periodically influenced by the dominating activities of the ovary.

The primary function of the ovaries is to produce the ova or eggs. These are enclosed in little bags or sacs called follicles, which begin by being very small and then gradually increase in size as they approach maturity, and in the last stages of ripening come to protrude visibly from the surface of the organ. When the cow is "on heat" or "in use" one (rarely two or more) of these follicles ruptures and discharges the ripe egg along with the liquid contents of the follicle into the body cavity. This act, namely the rupture of the follicle and the discharge of the ovum, is known as ovulation. The ova are microscopic in size or only just large enough to be detected with difficulty by the naked eye. Each ovum shed from the ovary in the way described is normally caught up in the trumpet-shaped end of the oviduct, which is a fine, somewhat coiled tube leading on each side of the body cavity from a position close to the ovary to the corresponding horn of the uterus or womb. The two horns of the uterus unite together further back to form the body of the uterus; the latter communicates posteriorly through a narrowed neck with the much larger passage, the vagina, through which the calf is carried in the act of parturition, and through a portion of which the urine also passes after being evacuated by the bladder.

After ovulation, which as we have seen takes place when the cow is "on heat" (i.e. during oestrus), the ripe egg passes down the oviduct. If the cow has been served the uterus and oviduct will as a rule contain very numerous spermatozoa or reproductive cells coming from the testicles of the male; nevertheless in the fertilization of one ovum only one spermatozoon participates, the rest of the spermatozoa dying in the female passages, their cell substance becoming disintegrated and eventually absorbed or got rid of. Similarly if the ovum be not fertilized it also dies and the substance of which it is formed breaks up and becomes lost. The union of the ovum and the spermatozoon is the really essential act in the reproductive process, the ovum becoming thereby endowed with a new vitality whereby it is rendered capable of undergoing those wonderful developmental processes which culminate in the full formation of a completely formed individual similar to those from which the ovum and spermatozoon were derived.

But as already mentioned, the ovary, in addition to the primary function of producing the female reproductive cells or eggs, has the further function of controlling the oestrous or sexual cycle. This it does by means of certain chemical substances or internal secretions which the ovary elaborates and which are passed into the circulating blood and carried to all parts of the body. These internal secretions which vary in quantity and in composition at the successive stages of the cycle act upon the mammary glands, the uterus and other sexual organs and excite or call into operation the periodic growth and activity which these organs from time to time display. The ovary is continually engaged in manufacturing these secretions; if the animal be spayed before sexual maturity the accessory organs (uterus, mammary glands, etc.) fail to develop; if the ovaries be removed after maturity the uterus and other organs cease to function and undergo atrophy.

For some time prior to oestrus the ovaries show an increased activity; this is manifested partly in the growth of the vesicles which contain the ova, but there is every evidence that the internal secretions of the ovaries are also produced in greater quantity. Thus the uterus undergoes growth, and eventually passes through a period of congestion which characterizes the period of "heat." During the latter part of this period ovulation takes place, and then the ovary undergoes a remarkable change. The vesicle from which the egg was discharged instead of merely shrivelling up as one might reasonably expect to happen, becomes transformed into a large yellow body which is highly congested, and comes to protrude from the surface of the ovary in much the same kind of way as the original follicle did before ovulation. But the yellow body instead of containing fluid is filled with tissue. The fate of this yellow body (which is called technically the corpus luteum) depends upon whether or not the cow has become pregnant. The presence of the foetus in the uterus reacts upon the corpus luteum and causes it to persist probably throughout the whole of pregnancy, and possibly for some time beyond. If the cow does not become pregnant the corpus luteum only lasts for two or three weeks, and is in a state of degeneration before the return of oestrus. These statements, however, are based rather upon an analogy with other animals than upon direct observations upon cattle, for which animals the records are all too scanty and very insufficiently worked out.

Recent research in all those animals which have been investigated has proved beyond doubt that the corpus luteum is a very important organ, for not only is its period of duration closely connected with the sexual rhythm, but the growth of the milk tissue, and consequently the mammary function, is dependent upon it. There is clear evidence that the corpus luteum, apart from the rest of the ovarian tissue, elaborates an internal secretion which passing into the blood reaches the cells of the mammary glands, and directly stimulates these cells to undergo growth and multiplication. Moreover the presence of a fully developed corpus luteum in one or both of the ovaries is correlated with an inhibition on the part of these organs to produce ripe eggs. And so long as a functionally active corpus luteum is existent in either ovary, oestrus cannot normally be experienced, and ovulation cannot occur. It is clear, therefore, that the period of duration of the corpus luteum is one of the main factors in the recurrence of the short cycle which in the cow is normally about three weeks.

The period of persistence of the corpus luteum is known to vary widely in different species of animals. It also varies in different individuals, and, according to Williams[1] the well-known American veterinarian, an hypertrophied or persistent corpus luteum may be a cause of long continued sterility. Why the corpus luteum should persist is not clear, but Williams states

[1] Williams, *Veterinary Obstetrics*, New York, 1909.

that this condition is most frequent in young cows and heifers which have suffered some time previously from contagious granular vaginitis. In certain cows oestrus is said to occur in spite of there being a corpus luteum present, but when this is so, the periods are shortened to 12 or 15 hours instead of the normal duration which is about double that time, and the cows if served in this condition fail to conceive. It is probable that in spite of oestrus taking place, ovulation fails to occur. The continuance of the corpus luteum as an apparent result of vaginitis or venereal disease may be due to the irritation of the generative organs, and is perhaps comparable to the normal stimulus set up by pregnancy which results in the persistence of the corpus luteum. The whole subject is, however, still obscure, and is in urgent need of investigation, particularly in view of the close connexion between the corpus luteum and the milk glands. Williams states that sterility as a result of persistent corpus luteum can be cured by elimination of the yellow bodies, and recommends that these should be squeezed out, the operation being performed either *per rectum* or *per vaginam*; and that oestrus promptly reappears within a short time of the corpus luteum being dislodged. According to the clinical records of the New York State Veterinary College at Cornell University, 95 per cent. of the cattle so treated conceived at the first subsequent service.

The recurrence of the oestrous cycle in cattle as well as in other domestic animals is also controlled by the condition of the animals. It is notorious that over-feeding results in sterility which is generally merely temporary, but if the practice be continued over a long period the sterility may become permanent. The custom of feeding animals so as to bring them into a high condition for show or sale is chiefly responsible for this unfortunate effect. Such practice, though it doubtless improves the superficial appearance, cannot be too strongly deprecated; nevertheless under the system now in vogue it is directly encouraged, since beasts with a good deposit of fat are more likely to become prize winners or to command high prices at sales.

Observation has shown that the reproductive organs of very fat animals may contain adipose tissue in no slight amount. In the ovaries of fat cattle this tissue acts in the same kind of way as the persistent corpus luteum. It is even possible that the condition of "fatness" may be a determining factor in the duration of the corpus luteum. However this may be the presence of fat in the interstitial cells of the ovaries associated with a great quantity of adipose tissue around the internal generative organs is a marked feature of fat heifers and cows. The interstitial cells of the ovary are believed to be the seat of formation of the internal secretion which is responsible for the recurrence of the heat periods. It is not surprising, therefore, that the deposition of fat in these cells should interfere with the elaboration of the secretion, and should bring about an inhibition of the normal cyclical changes.

In addition to the fatty deposit in the ovarian interstitial cells, the follicles or vesicles which contain the ova are also affected as well as the ova themselves. The ovarian follicles in the normal animal are in various stages of development. It is probable that all the ova are already formed even as early as birth; during growth they undergo a process of gradual development, the follicles which contain them likewise growing until as sexual maturity is approached each vesicle becomes enlarged and filled with a nutritive fluid. At the first heat period one or more of these vesicles is so large as visibly to protrude from the surface of the ovary, and eventually, at the time when the animal is ready for service, ovulation occurs, the vesicle undergoing rupture, and the ovum contained within it being discharged. This gradual process of ripening takes place throughout the whole period of reproductive activity, successive batches of ripe follicles and ova being matured preparatory to ovulation at every recurrence of oestrus. It follows that the ovaries at all times contain follicles in various degrees of ripeness, and that the rate at which these mature and the number which are ready in any one sexual season must depend upon the nourishment supplied to the developing eggs at previous stages in their life-history. Any interference with the supply of the right kind of nutriment whether by the animal being too fat or by her having been for a prolonged period underfed must affect the supply of ova and may induce a condition of sterility.

Observations upon the ovaries of young cows and heifers which have been excessively fattened have shown that the ovarian follicles and their contained ova are liable to undergo a process of degeneration. The extent to which this occurs varies with the condition of fatness. It probably seldom happens that all the ova die in the ovaries, for the less mature ones are not so easily affected as those which are well advanced towards maturity. It follows that an animal which had been rendered sterile for a whole season may recover in the next season, and produce ripe eggs, which can be fertilized by spermatozoa. If, however, the deposition of fat had been carried further, the fertility may be impaired in subsequent seasons owing to a failure in the supply of sufficiently mature ova. Thus animals which have been got into "high" or "extra" condition in response to the demands of the judges at shows, are liable to remain uncertain breeders for some time subsequently, and while they may or may not be able to perform the sexual act, are unable to conceive. It is unfortunate that the prejudices of the show and sale-ring in favour of very fat animals should be attended by consequences which may permanently lower their reproductive efficiency. To quote the words of Cornevin: "Chaque année, dans les concours, nous avons sous les yeux des spécimens des plus belles races....qui, véritables modèles de bonne conformation, de puissance assimilatrice et d'aptitude à prendre la graisse, restent stériles[1]."

The condition most favourable for reproductive activity is a good thriving one, neither too fat nor too lean, and with a tendency to improve rather than to go back. This fact is well illustrated by the practice of "flushing" ewes, that is supplying them with additional or improved feeding whether corn, cake, or turnips or superior pasture at the approach of the tupping season. By such means not only is the fecundity improved, the number of twins produced at lambing being perceptibly increased, but the periods of oestrus are hastened forward and the ewes got in lamb at an earlier date. By analogy with sheep it ought to be possible to discover appropriate methods of feeding and treatment by which cows and heifers can be induced

[1] Cornevin, *Traité de Zootechnie*, Paris, 1891.

to come in use in the winter months and so produce calves and subsequently lactate at a season when under present conditions parturition is infrequent. It must be remembered, however, that the heat periods of the late autumn are sometimes very transient and hard to recognize in cattle, that it is often only by close and constant observation and trial that it can be discovered that a cow or heifer is in a condition to take the bull, and consequently a "cowman" may let a "period" pass, and perhaps lose the chance of getting the cow in calf for that season, through failing to detect any vaginal discharge or other sign of oestrus.

In this chapter reference has been confined to certain problems in "agricultural physiology" of which the writer has had direct cognisance or practical experience. In the preface to this book attention has been directed to the need for investigation upon heredity and the way in which this can best be done with the view of obtaining practical results. The other branches of animal physiology contain fields of inquiry of equal importance, both for the agriculturist and for the man of pure science, and innumerable instances might be given. But enough has been said to indicate the nature of the work, the high interest which the subject possesses, and the need for applying scientific method to the live-stock industry, not only for the benefit of those directly concerned with it, but on behalf of the nation as a whole.

CHAPTER X

BREEDS OF CATTLE

So much has already been written by authoritative writers about our British breeds of horned stock, that this chapter is bound to cover a certain amount of old ground. To omit it entirely, however, would be to miss the opportunity of recording the result of 20 years' study and observation of the possibilities of improvement in our national herds. The information on which conclusions have been based was not always exact; instructors were frequently prejudiced in their views by affection for some particular breed; many, alas, of those who supplied information had only their own profit in view when discussing the merits of a particular variety. Yet in spite of these and many other disadvantages one may still hope that the information collected may be of some use in directing future work. Knowledge would be much more accessible to the student if it were more extensively standardized. At the present moment such things as weight, early maturity, constitutional vitality, milk-yield, quality of flesh, horn, skin, "type," even description of colour, have no precise meaning to the majority of cattle-owners, even among those who have a considerable part of their capital invested in such stock. Nomenclature is so vague that one often has difficulty in understanding its exact meaning, and *a fortiori* in conveying it to others. Hence it is difficult to write about pedigree cattle; but as they are for many years to come likely to be our only means of grading up non-pedigree animals, the attempt must be made.

Perhaps a few rough definitions will help to introduce some sort of order into the descriptions of the different herds, to make the meaning clearer, and to avoid tedious repetition.

As regards size, it may be said that, when "large" animals are referred to, the following approximate weights are meant: cows at 5 years or more in good breeding condition will weigh about 1,400 lb., when moderately fat some 200 lb. more;

bulls 4 years old and over in good store condition will weigh quite a ton, when fat 2,400 lb. "Small" animals in the same condition and of the same age would mean females weighing 1,000 lb. and males 1,500 lb. "Very large" exceed the former and "very small" fall short of the latter figures. Between these weights cattle will be found described as "medium sized." Obviously, weight is not the only factor which determines size, and frequently some qualifying adjective must be added.

"Strong bone" denotes an animal well furnished with weighty limbs and implies that the proportion of bone in the skeleton generally is great compared with the flesh it carries. An animal is said to be "coarse" not only when it is strong-boned, but also when the excess of bone prevails at those parts where it is least desirable. It is said to be "deep" when the depth through the body is great compared with the length of limb; in the same way it is said to be "leggy" when the reverse is the case. The expression "bony," on the other hand, should be used to denote absence of an adequate supply of muscle and other tissue wherewith to cover the frame; a "bony" animal need not be "strong-boned." The reverse of "bony" is "thick-fleshed"; an animal may be strong-boned and thick-fleshed; if it is strong-boned and "thin-fleshed" or fine-boned and "thin-fleshed," it will in both cases be rightly described as "bony." "Quality," used without any qualifying adjective, generally means a fine-boned animal showing no coarseness, and, when applied to specimens of any one beef breed, further denotes that its covering of muscle and other tissue is ample, that is to say, it is "thick-fleshed." The word "neat" is used to denote that an animal is symmetrical and well put together. It implies that the various parts of the body are well proportioned in length, that the bone is the reverse of strong (generally fine in fact), and that the absence of all angular projections is a marked feature in its appearance. If a beef animal is "neat," it is largely due to a thick, regular covering of flesh on a well-sculptured frame; in a milch-cow or one of a purely dairy breed (that is to say, useful for butter or cheese-making purposes) the feature is due to particularly fine bone being draped with a soft, pliable hide.

In referring to the characteristics or qualities looked for in any particular breed, the word "type" is often used, coupled with "breed," to denote the possession of such qualities; "character" is very frequently used as synonymous with "breed-type."

"Touch" is another expression that requires defining. It is largely dependent upon the skin. A good skin is elastic and of the proper thickness; the good beef animal is generally supposed to carry a moderately thick, the dairy cow a moderately thin, skin. Experience leads me to think that the various thicknesses of hide are among those characters which indicate the two purposes, i.e. meat and milk; good skin, however, is only part of the perfection; for under it one expects to handle springy tissue; further, the hide should glide easily over the underlying substance. If the animal is being felt over by the hand for its beef-yielding quality, the "touch" should denote a thick covering of firm yet resilient substance in the space between all parts of the skeleton and the skin; in dairy stock one does not look for the same thick lining but, on the other hand, the dairy cow has not a good "touch" if one feels nothing but hard bone under her skin.

The Shorthorn.

This breed may be said to be spread wherever the English tongue is spoken. In colour it is red, white, red and white, or roan; the last-named tint has many variations—from a deep red just flecked with white to a white skin just darkened at the head, neck, and other extremities; indeed, some very beautiful roans also show a tinge of blue. Though perhaps not so variable as colour, other characteristics of the breed differ very considerably. Breed-type or "Shorthorn character" seems, however, to have been so definitely kept in view by all breeders that large or medium sized, red or white, thick-fleshed or lean, these cattle may be recognized by anyone with a claim to knowledge of stock.

I take the following description of this celebrated breed from authorities quoted by Sinclair:

"The breed is distinguished by its symmetrical proportions,

and by its great bulk on a comparatively small frame; the offal being very light, and the limbs small and fine. The head is expressive; being rather broad across the forehead, tapering gracefully below the eyes to an open nostril and fine flesh coloured muzzle. The eyes are bright, prominent, and of a particularly placid, sweet expression; the whole countenance being remarkably gentle. The horns (whence comes the name) are, by comparison with earlier breeds, unusually short. They should spring well from the head, with a graceful downward curl, and are of a creamy white or yellowish colour; the ears being fine, erect, and hairy. The neck should be moderately thick (muscular in the male), and set straight and well into the shoulders. These, when viewed in front, are wide, showing thickness through the heart; the breast coming well forward, and the fore legs standing short and wide apart. The back, among the higher bred animals, is remarkably broad and flat; the ribs, barrel-like, spring well out of it, and with little space between them and the hip bones, which should be soft and well covered. The hind-quarters are long and well filled in, the tail being set square upon them; the thighs meet low down, forming the full and deep twist; the flank should be deep, so as partially to cover the udder, which should not be too large, but placed forward, the teats being well-formed and square-set, and of a medium size; the hind legs should be very short and stand wide and quite straight to the ground. The general appearance should show even outlines. The whole body is covered with long, soft hair, there frequently being a fine undercoat; and this hair is of the most pleasing variety of colour, from a soft creamy white to a full deep red. Occasionally the animal is red and white; the white being found principally on the forehead, underneath the belly, and a few spots on the hind-quarters and legs; in another group the body is nearly white, with the neck and head partially covered with roan; whilst in a third type the entire body is most beautifully variegated, of a rich deep purple or plum-coloured hue. On touching the beef points, the skin is found to be soft and mellow, as if lying on a soft cushion. In animals thin in condition a kind of inner skin is felt, which is the 'quality' or 'handling' indicative of the great fattening propensities for which the breed is so famous."

In the above description the late and much revered Mr John Thornton is decidedly inclined to favour, by his description, the type of animal that is better for beef than for the pail. I will, however, refer to this matter in detail later in this chapter.

The head of any high grade animal is always considered of the utmost importance and this is particularly the case with the Shorthorn. The best description of this vital point that I have read is the following[1]:

"To describe the modern Shorthorn we begin with the female, and take first that index to breed and character—the head. In the highest type of head the face shortish, broad across the eyes and forehead, generally a little hollow in the outline of the face, and decidedly so between the eyes, finely cut out, like artistic carving in wood or stone, down the face and round the muzzle; the nostrils large and open. Cheeks not too fleshy[2], eyes bright but placid, horns wide set, and somewhat flat at the roots, growing outward at first, and as the animal advances in age forming gentle curves which should not be immediately upward.—The horns and the muzzle should be light coloured and clear, the muzzle a palish buff without stain of black..."

The remarkable thing about the Shorthorn breed is the eager, constant, and universal demand that has continued for over half a century for sires of this variety as improvers of herds *wanted for beef-production*. This is all the more astonishing when one reflects that the breed was created to be of dual-purpose capacity, that it is capable of producing meat and milk for its owner at the present day (provided that care has been taken to select

[1] *Cattle, Breeds and Management*, by William Housman, 1900, Live Stock Handbooks, No. 4, Vinton and Co.

[2] It surprises me that neither of the above quoted authorities refers specially to the jaw. This should be deep from eye to "angle of the jaw," should be clearly defined (i.e. not lost in loose folds of skin) and be well turned and covered with firm muscles that should stand out well. Many authorities look upon this point as *very important*; and, considering the effect that mastication has upon the food, they may be right.

breeding stock with both qualifications), and that other breeds selected for very many generations with the sole object of beef-production have never been serious rivals for the bids of those wanting to breed large numbers for the shambles. These reflections cannot but win our respect for the originator of the "Improved Shorthorned Breed"; they make us consider with some anxiety the possibility of systematic or scientific effort in the twentieth century doing anything to improve the empirical methods of the eighteenth.

Many explanations occur to one's mind of the phenomenal success of the Shorthorn in securing the custom of so many of the buyers from overseas.

Without doubt one of the most valuable characteristics of the breed is its adaptability. It will thrive under many variations of climate, soil, and other conditions. After a few generations of selection it will direct the metabolism of its system towards beef-production at the expense of milk-production. The original red colour of its coat, we are told, was a pale red inclined to yellow. Following the fashion of overseas customers, a deep blood-red colour is now common. To suit pasturage that is very productive a large beast can be secured; on the other hand, medium, or even small, cattle can always be produced from Shorthorns when required. In a harsh climate the breed can be relied upon to grow an abundance of soft, warm, weather-defying coat; in more favoured climates this faculty is not brought into play. Given plenty of food to force growth, well-bred stock may be relied upon for very early maturity; on the other hand, when "done" only moderately well, the animal will gradually and slowly mature. No other breed can show such adaptability.

It also has a very great reputation for "prepotency" as a sire. Breeders believe that it has the faculty, developed to a very high degree, of impressing certain good qualities upon its off-spring. The possibility of any race of animals having this pro-perty developed to a greater extent than another has been questioned of late years. It is, however, very firmly believed by many of the most experienced that the Shorthorn has it to a very marked degree; I, personally, am very much inclined to believe that the view is correct. It is quite certain that the first crosses from a Shorthorn bull mated with nondescript females show to a marked extent the good qualities of their sire. If the cross is repeated through a few generations, the good qualities are reproduced regularly, or, as it is described, the stock breeds true. The transformation is specially marked when the foundation female stock are thin-fleshed, coarse, or slow-growing, or have these bad qualities in combination. It is to be noted carefully that we are dealing here with fleshing qualities. The power of imparting deep-milking has not been so carefully recorded and observed in the breed as to warrant the expression of any very definite opinion.

The single characteristic to which I am most inclined to attribute its pre-eminence in demand as a sire of beef-cattle is one to which my attention was drawn many years ago. This peculiarity has no name, but it is easy to explain. A Shorthorn, or an animal in which the Shorthorn blood prevails, is fit for slaughter with little preparation. That is to say, oxen of this breed furnish useful carcases for the butcher without as much special feeding as is required by animals of many other varieties. This is an attribute of which the value cannot be over-estimated from the point of view of many foreign buyers. A breeder owning a very large herd, or a great number of herds in a climate liable to long periods of drought, would be very much handicapped were he forced to keep all his saleable stock till they were prime. If the breed being grazed had all to be held over till they were quite finished, he would run the risk, in years of drought, of finding the whole of his capital tied up without bringing in any income at all. Shorthorns, on the other hand, are quite fit for the butcher once they are at that stage which is called "meat," as contrasted with the condition known as "flesh and bone," and need not be kept till they are "prime." Through this idiosyncrasy of the breed an owner of vast numbers may always rely upon having a certain proportion of his stock in a condition that enables him to cash them, and so upon having money available for current expenses.

The origins of this characteristic are well worth scientific investigation. It occurred to me several years ago that the reason why two-and-a-half year old[1] Shorthorns were fit for slaughter at a leaner stage than most other animals was due to the fact that their flesh was "marbled" at an earlier stage in the feeding period; I have been greatly strengthened in this belief by the rough observations I have been able to make during several years past. At the same time, the matter is so much open to doubt and the results, whatever the cause, are so important that it is a clear case for systematic research.

For reasons already suggested, it is difficult to do more than quote such a description of the characteristics of the Shorthorn as has already been given. It will be well, however, to refer to some economic characters; and these must be treated under two heads, namely those of the beef, and the dual-purpose, members of the breed respectively.

The purely beef animal is the product of the demand from overseas. It is a breed that has been evolved out of the tribes founded by the late Amos Cruickshank, and to a lesser extent by his brother, both Scotsmen from Aberdeenshire. It was my good fortune to be told once by their greatest English disciple, a man who had had the advantage of frequent interviews with Amos, what was the belief of this celebrated man. "Amos Cruickshank told me," said this great breeder, "that the great thing to look for in a Shorthorn was *constitution*." I gathered that by the word *constitution* was meant an animal with a particularly deep body, with wide well-sprung ribs, those at its fore-end or "over the heart" being specially well arched, with a very capacious abdomen, the muscles of whose walls were strong enough to contain it in such a way as to avoid any appearance even of looseness or "paunchiness," and, finally, with a skin well coated with abundant, soft, yet weather-defying hair[2]. These points have been well maintained—without

---

[1] I am not by any means convinced that the faculty is more pronounced in very young Shorthorns than in baby-beef animals of other descriptions.

[2] I think it may be of interest to record that I submitted what I have here written to the celebrated breeder I refer to. This gentleman—Mr J. Deane Willis of Babton Manor, Wilts.—writes (6. i. 19): "You are welcome to say I agree about constitution. It was after I bought all his yearling heifers in 1889 that I used to go up and see him [Amos Cruickshank]."

flattery one may say that they have been developed— by present-day breeders.   Amos Cruickshank's ideas *did not*[1], however, involve the total neglect of milking quality in his stock; this has been referred to earlier in this work, and so the neglect of later years need not be emphasized here.

The present money-value of the best beef Shorthorns is influenced by colour to a surprising extent.   A deep rich whole-coloured red, or a very evenly distributed dark roan, is saleable at quite a different figure from a yellow-red, or a red and white pied.   A whole white, on the other hand, will sometimes sell for a very high price to mate with bad coloured reds and red and whites, for white animals when so mated have a reputation for breeding good roans.   White legs on any coloured specimen are altogether anathema.   One would imagine that these peculiarities about colour are merely foibles of the show-ring, but it is not so, for the fashion for certain colours is set up by customers who are buying for commercial purposes.   It cannot be easily believed that a red and white sire will get any worse breeding stock than a dark rich roan, and yet the man who wants to breed bulls to turn in among a thousand cows used for nothing but beef-production will pay a far higher price for the animal showing the latter colour.   On the other hand, it is not altogether wise to attribute the preference to prejudice alone.   For instance, there is a great prejudice among the members of the cattle industry in England against white colour.   This may be pure prejudice; yet you will find veterinarians of experience talking of "white-heifer disease," a rare derangement of the reproductive organs which, these practitioners will tell you, is nearly always found among white, or very light-coloured roans. Certainly, cattle of particular colours have the appearance of being better furnished with flesh than animals of other hues. Whites are always said to look larger than they really are— I myself think dark rich roans look well-fleshed—but ideas of this sort would hardly be important enough to create a demand, making a great difference, often amounting to hundreds of

---

[1] Two writers of the present day, Sinclair and Robert Bruce, uphold this. See also article by W. Housman in the Royal Agricultural Society's *Journal* for 1880, p. 384.

pounds, in the value of one stock bull. In Shorthorns, I personally have noticed that the widespread belief that a "yellow-red" is a better colour, as an indicator of milk, than a "blood-red" has a considerable foundation in truth.

As a breed, Shorthorns generally may be classed as large animals. In the beef classes size on the whole is in request. That is to say, of two animals having much the same type and showing approximately the same good beef qualities the larger would win, and it may be said that the market in this respect emphatically follows the show-ring. A medium-sized animal of great merit would, however, always be graded above a very much larger specimen showing bad points by anyone fit to be a judge of Shorthorns.

No impartial critic can claim perfection for the Shorthorn as a beef-breed. His greatest fault in conformation (especially in the class often misnamed "Scotch") is at the quarters—that part of the carcase from which the butcher cuts the "rump-steak." Here the Shorthorn is apt to fail, relatively to his other good points; either the meat is not thick enough or there is too large a proportion of fat to lean, or the animal is apt to be coarse. More especially is the excess of bone liable to be shown at the tail-head. Probably his worst feature as a beef-beast is his propensity to carry too large a proportion of soft fat, his flesh being often the reverse of firm under hand all over the body, and, further, he often carries a lot of superfluous or "patchy" fat more suited to the tallow-tub than the carving-dish. The foreigner demands that the stock he buys should be "well up"; in other words he likes stock suffering from an excess of obesity which is positively pathological. The show-world has responded so faithfully and so long to this demand that every possible encouragement has been given to the breeding of animals that have a tendency to grow too much fat. It is common to hear, or even to read, the remark that a certain animal has "stood training well"; this means, in plain English, that it has stood over-feeding for a long time without becoming "patchy" or actually falling away through ill-health.

Both types of Shorthorns are very good as economic consumers of feeding-stuffs. For their size (and it will be remembered that they vary very greatly in this respect), I have always looked upon them as easily satisfied as regards quantities of bulky foods. One is forced, however, to admit that there is no reliable evidence to support this. It is one of those matters which require to be tested by carefully-conducted and extended feeding trials. On the other hand, I think it must be admitted that Shorthorns require more consideration as regards quality of bulky fodder than several other breeds. I have seen many instances of other cattle apparently doing better than Shorthorns on a diet of straw and rough coarse hay, or, again, when grazing on weedy, rank pastures. They do not, particularly when in very high condition, stand temporary shortage of pasture especially well; and though they can, as a breed, adapt themselves to living on moderately short commons, other breeds have a reputation for doing so better. While they will thrive in very many climates, it seems that they suffer more than the very hardy breeds when exposed to extreme wet and cold; it must not, however, be deduced from this that the breed is in any way delicate. Large numbers of the variety undoubtedly are found to suffer from tuberculosis when submitted to the tuberculin test. This I believe to be due to their being so often kept under insanitary conditions, and of course it may be that the number which react is not relatively large; but the fact that many react, while apparently few suffer signs of real ill-health from the disease, has always appeared to me to show that among their other good qualities they have considerable constitutional vitality with which successfully to combat the evil.

Turning now to the dual-purpose Shorthorn, we may say at once they are not so massive, or blocky, as their purely beef-bred fellows. The lines on which they are built are finer, their ribs, especially over their hearts, are not so well sprung, their thighs do not carry so much flesh and look "better bent," they are lighter and look longer in the neck, and their heads are decidedly longer from eye to nostril. They should, however, be quite as deep in the body, their ribs next to the loin should be as well sprung and wide, and there should be as great width between the eyes. Their skin should be thinner, but not "parchment-like"—in fact to be what is well called "papery"

is the very antithesis of the true Shorthorn type. The Shorthorn that belongs to a deep-milking family is not expected to
be as thick-fleshed as those belonging to the other branch of
the breed; she is apt to be somewhat too fine and may in fact,
if care is not discreetly employed to avoid it, become as bony
as one of the breeds famous for butter-making purposes without
giving anything like the same amount of cream in her milk as
such animals do. Further, *if care is not taken in mating these
"skeleton milkers"* one may breed cows having the bone and
thin flesh of one grandparent combined with the absence of
milk-yielding power belonging to another ancestor.

The udder in this breed, for many of its number are known
to be failures at the pail, is of particular importance. It should
be carried well up and back between the hind legs, and should
be large in size but not too pendulous. I have not found,
personally, that a "globe-shaped" bag, equidistant teats, and
other fancy show-points are special indications of high milk-
yield. I have seen so many splendid milkers with bags absolutely
wanting in a number of the attributes demanded by the show-
ring that I am inclined to believe that systematic investigation
would prove that these attributes have no great intrinsic value.
Ample size and great quality of touch are the vital matters.
The importance of touch cannot be over-estimated. The skin
of the udder should be moderately thick only, there should be
plenty of it, and it should be elastic. If a fold of the skin be
taken between the finger and thumb, the two halves of the
plait should glide over one another in a manner pleasing to the
hand, the hair should be silky, or at any rate not thick and
harsh, and one likes to see a net-work of small veins under the
hair when the udder is distended. The gland under the skin
should touch firm, but elastic, having no sort of tendency to
feel doughy or soft and fat. What is known as a "fleshy" bag
is often large, but yields little milk.

Besides the udder itself the veins which run along the lower
parts of the abdomen should be specially noticed; these vessels,
when the cow is in profit, should be large enough to stand out
and be easily seen; they should handle firm, they cannot, it is
said, be too contorted, and the holes through which they

traverse the bony walls of the under part of the body cavity *should be large*. These holes, which are often called "wells," are a valuable help when examining a *dry* cow.

As regards amount of yield, no cow of this breed should be looked upon as good if she does not average 700 gallons a year for several periods of lactation succeeding the dropping of her third calf. With careful breeding a herd may, without great difficulty, be got together in which the cows will average 700 gallons for the first six years following the first calving. It must always, however, be borne in mind that this cannot be expected unless a very careful milk-record is kept. This is of great importance in all breeds, but with the Shorthorns it is imperative, for it is quite easy to find specimens of the breed very thin-fleshed indeed, yet giving no more milk than those whose flesh-forming capacity has been developed to the greatest possible degree. It is no exaggeration to say that it is easy to breed these specimens among the best strains; their common occurrence in our herds may be noted by any unprejudiced observer.

The milking qualities of the cattle can easily be kept up to standard if high records to the credit of all dams used for reproduction are resolutely insisted upon; it need not again be emphasized that this is even more important in the case of a bull's female progenitors than of a cow's.

The problem of systematically maintaining the inheritance of beef-production on the part of the milch-cows and their calves has still to be solved. In the past, selection for this capacity has been left entirely to the judgment of the owner. That this judgment has been remarkably good in the case of individual breeders cannot be held to be evidence in favour of continuing such rule-of-thumb methods; for typical examples of Shorthorn misfits bear witness in almost every market-place to the folly of unsystematic breeding.

The improvement of meat-production qualities among all milking Shorthorns can and ought to be immediately taken in hand, so that the fine specimens which are now unfortunately the exceptions may become the rule. The weight for age, the measurement necessary to secure quality, can be taken at once as a basis for the establishment of a beef standard. This beef standard should then be made a further qualification which, besides a milk-record pedigree, should be added to the certificate of any animal used to reproduce its kind. Possibly, nay probably, such work done in the interests of a beef standard would enlighten us on many points now obscure, and might give us indications, other than weight for age and measurement, that ought to be included in the standard. An obvious danger is that of excessive feeding to obtain weight for age. It seems very probable, indeed, that very high development of adipose tissue may, in the case of young heifers, interfere, by infiltration or by otherwise inhibiting the development of mammary tissue, with the power of lactation in later life. Even with young bulls this condition may be harmful. I know of one young Shorthorn bull, with a pedigree exceptionally rich in milk-yielding females, who proved sterile, and whose reproductive organs were found on post-mortem examination to have degenerated through excess of fat among the tissues of the testicles. But this and other similar difficulties that may occur will not be found insuperable, if only the breeders will lend their present intuitive skill to help scientific work in the future.

That such help given by the experienced Shorthorn breeder would be of the highest value to the country, I firmly believe; for, if increased fruitfulness of our land is to be effected concurrently with prime beef-production, the one benefiting the other, there is no variety likely to be so sure a foundation on which to build improvement as the famous " Red, White and Roan " cattle which the disciples of Bakewell created in the latter half of the eighteenth century.

Lincolnshire Red Shorthorns.

The Shorthorn, when registered, has his pedigree inscribed in Coates's Herd Book, founded in 1822. The Lincolnshire breeders of cattle much the same in character, have for the last 23 years started a herd book of their own.

There is, however, a more material distinction than mere registration between the Lincoln Red and the Coates's Herd Book cattle. The Shorthorn, the old variety, is large, but the

Lincoln Red is larger; it is stronger boned, and perhaps rather less well sprung in the rib. Red, as the name implies, is the only recognized colour; though animals with some little white markings and even roans appear from time to time among stock that have been registered for several generations in the Lincoln book. The type is also rather different and not so specific in character. Lincoln Red breeders require a character in the bulls which I have heard them describe as "sour," whereas the "Shorthorn man" would be shocked and pained at his stock being so described. The "sour" animals are less fine about the head, of which the lines are less finely chiselled; there is slightly more skin at the throat and the whole of this part of the body has a more massive appearance. Such a head coupled with a body carrying more bone especially above the hock and below the knee, all belonging to a very large animal, briefly constitute the ideal of the breeder who is not frankly trying to reproduce the Coates's Herd Book type.

The Lincoln Red can also be divided into two groups—beef and dual-purpose cattle. Among the latter are to be found some very deep-milkers indeed. They have the reputation of being very hardy cattle and of thriving on coarse fodder and pasture. In my own experience of the breed I have always thought that they were cattle requiring a great deal of food, but that their digestions and appetites enabled them to be less particular than some others as regards quality of provender. They are, when forced, quite up to the average among breeds celebrated for early maturity; on the other hand, if not exceptionally well done, they lack quality, and so are seldom slaughtered as baby-beef. Some graziers and winter feeders proclaim the Lincoln Red to be the best bullock *for the farmer*, though the evidence to uphold this claim has hitherto escaped my notice. The butcher's objection to their excess of bone has, on the other hand, seemed to me very clear in the case of many specimens I have observed being sold on public markets.

CHAPTER XI

The Hereford.

This breed is very generally supposed to belong to the group of beef-producing, and nothing but beef-producing, animals. It is certain that a great many of the cows have little to spare after nursing one calf; some fail, to a certain extent, even in this. Among animals known to be pure-bred some look much more likely to milk than others, but I have never seen a specimen with a really high milk record to her credit; but, as is so often the case with beef-breeds, records are hardly ever taken. Indeed, some owners of pedigree herds are said to object to a cow that shows anything approaching a fine, big, productive-looking udder after calving! It is therefore in relation to its reputed single capacity of beef-production that we must consider it here.

A large animal, red, with a white face and fine massive pale-coloured horns, it is always a pleasing feature both in the county from which it is named and in the beautiful Severn country generally. An expert on breed-points alone can prescribe the right tint of red, the proper white body-markings, and other indications of good Hereford character; but animals of the breed will bear analysis as fine specimens of the beef-beast without details of this kind. It is often looked upon as the perfect animal for what the stock-breeder calls "meeting you well." The fore-end of a beef-beast should show quality to the greatest possible extent, the skin of the neck should cover a short, thick wedge of solid meat, and there should be great width at the breast, i.e. from one point of the shoulder to the other. This width is not obtained through the largeness of the bones of the "shoulder-points." Width gained by large bony "points" constitutes coarseness, which is the reverse of what the good Hereford shows at this part of his body. He is wide below the breast right away down to the floor of his chest, and the "dewlap," as the whole of the anterior of the *sternum*, or breast-bone and its covering, is called, projects well beyond the line of his well-muscled, or "meated," forearm. All this gives the beast the massive and weighty appearance when viewed obliquely from the front which is summed up in saying that he "meets you well." The rest of his body in front of his hip-bones is good, but behind these (called "hooks" in the trade) he is apt to be less perfect than in his middle or at his fore-end.

The fault at the hind-quarters, often found in Hereford cattle is two-fold. The part of the body, between the ilium and ischium bones, that carries the meaty mass from which the butcher carves the rumpsteak, is sometimes cut away owing to the anterior portion of the spinal column dropping between the line of the hocks and the tail-end. This conformation must undoubtedly reduce to some extent the proportion that the rumpsteak piece bears to the whole carcase, and so rob the meat-salesman of some of this super-quality meat. Below the part from which this is cut and anterior to it is found the first quality joint known as the "round"; this, again, is not as well developed as is desirable owing to the meat not being carried down as close to the hock as it might be. Among show Herefords, especially the prize-winning bulls, these faults have largely disappeared of late years, but to a certain extent they are still common among those sold for feeding. Perhaps the word "fault" is too strong; "imperfection" is probably a better word to be applied to these well-shaped cattle, as compared with average cattle of other beef-breeds.

For grazing, in the restricted sense of grass-feeding, the Hereford has a very high reputation; some authorities claim that on an ordinary first-class pasture it will give better returns than any other breed. Here, again, there is no direct evidence either to support or to refute the claim. My own observation leads me to believe that Herefords will fatten rather more quickly on ordinary good grass than other breeds, but I should be surprised to find that the advantage was very appreciable; certainly I would not expect the superiority to be anything like as high as is claimed by some enthusiasts. The question of whether on the *very best finishing land* the Hereford would do any better than other good beef-breeds, is more difficult to answer. The faculty of doing very well upon grass just below the best is, to my mind, the greatest economic asset of the breed. The meat from **grass-fed** Hereford bullocks is as good in quality as that from average first-class beasts of other breeds; many authorities would call it superior to any other, and I know of no evidence to controvert such a view. To get fat on aftermath when allowed a *moderate* ration of concentrated food is, in my experience, a good quality peculiar to this breed for which the "white-faced" cattle do not receive adequate praise. Few, however, will admit that the Hereford fed in the winter on cake and roots and other plough-land food supplies the best quality meat. Many years ago my attention was drawn to severe criticism of the breed from this point of view, and there were, in particular, many complaints that the amount of "kidney-fat," highly valued in November and December on so-called Christmas bullocks, failed badly. Being much interested in the breed, I have watched for this fault very persistently, and have come to the conclusion that there is no ground for the complaint. I strongly suspect, however, that if on this yard-feeding a Hereford is produced ripe enough to carry an adequate mass of kidney-suet wanted for plum-pudding making, he will be found to be *excessively fat* all over the rest of his carcase. This fault is very much in evidence when the beast is handled alive, for many Herefords are apt to be soft to the touch.

Butchers have another complaint to make against him, namely, that he always looks bigger than he is. This is not a characteristic for which the farmer or grazier should find fault with the breed; but the trade has been known to speak of the Hereford as the "white-faced robber"! Our bad system of marketing and not the breed, ought in justice to be blamed for the irritation which produces this unjust complaint. There is evidence (slight, but nevertheless emphatic) to show that those accustomed to buy other cattle *by eye*, will, when they get the carcase on to the shop scales, be so much dismayed at their lack of skill in judging both live and dead-weight that they will admit, for once, that purchase *by weight* has some points in its favour.

The Hereford hide is particularly thick and yet very elastic to handle, and is often covered with a beautiful coat. The tanner shows the particular value he sets upon it, for not only is it very heavy, as one would expect from its thickness, but a special and higher quotation is always offered on the hide-markets for "Herefords."

As regards future beef-production in Great Britain (for the foreign demand is likely to keep up an active production in high-class herds bred for that trade) the breed promises to be useful along two distinct lines. In the first place, Herefords are admirable cattle in districts where, owing to difficulty of access with tillage-implements, the countryside may economically be left under permanent grass, even though the pasture is only second class. But there is a second purpose for which Hereford bulls are particularly suitable, but in many parts of England are not sufficiently used, namely, for crossing with the number-less cows, which, being thin-fleshed, are not good enough milkers to be used as progenitors of future milch-cows. Young cattle so bred make excellent stores for any class of feeding, including the production of baby-beef. Some rearers of store-stock are now wise enough to place a good bull at the service of the owners of large herds of such cows and to take their calves at a contract price; it would be well if such co-operation of interests could be systematically extended. Another good plan, at present only very occasionally adopted, is to mate a Hereford bull with all the heifers on a farm; these can all be tested for milk-yielding quality by the recording sheet and those proving deep-milkers can be subsequently mated with a bull of known milking blood. This, again, could be more widely and economically carried on, if co-operative methods were more prevalent.

The Devon.

This small ruby-red animal is famous for its wonderfully symmetrical and deep-fleshed carcase. It is seldom used for beef-production alone, for in nearly all *commercial* herds the cows, besides rearing the calves, yield a considerable, if not very large, quantity of milk from which comes the famous Devonshire clotted cream, the blue Dorset skim-milk cheese, and sometimes even new milk for sale in the towns. It remained, however, for Messrs A. and T. Loram, of Rosamundford near Exeter, to show the cattle-world that high-bred Devons could give a thousand gallons of milk in the year. There would seem to be little doubt that, with care, deep milk-yield could be added to the splendid fleshing properties of the breed. I specially examined Messrs Loram's stock from this point of view and could find no evidence of their falling-off as good beef-producers.

As a beef animal, it is hard to find fault with the North Devon. Here it must be explained that the breed is still divided into different types. The "North Devon" type is smaller, is of a richer red colour, appears to be thicker-fleshed and is better put together. The other type, which is somewhat larger and more sand-coloured than true red, is sometimes spoken of as the "Somerset Devon"; it must not be confused with the "South Devon" from which it is entirely different. Of late years there has been a tendency to get both types of Devons larger, and one is inclined to wonder whether this is not a great mistake. There is a very marked demand in our thickly-populated country for small joints of high-class meat. Now of the Devon's many good qualities as a butcher's beast the fulfilment of this particular demand always seems to me the greatest. For, having regard to the size of the joints, I know of no beef that shows such depth or thickness of lean meat as that cut from the carcase of the Devon. Beyond this I have little more to say of the Devon; he is good all over rather than at any particular point. Animals of the breed are reputed to be small eaters even after allowance is made for their being little cattle. I have never seen "beeflings" of this race, and indeed have heard that early maturity is not among their leading characteristics.

In the interests of the future the best qualities of this breed should be sifted out from their imperfections, for they undoubtedly are cattle worthy of all that research can do. There is not, to my knowledge, any breed in England that so completely keeps the Shorthorn out of the chief markets of the district from which it originates as the Devon.

The Sussex.

To-day the Sussex is essentially a beef-breed; one hundred years ago it was essentially the draught beast of the heavy Weald Clay in the county from which the name of the variety is derived. The transformation is one of which the Sussex breeders may be proud. For the working ox, of which the eighteenth century "Sussex" was a very typical specimen, was leggy, strong-boned and coarse, not particularly well sprung in the rib, slow-growing, and carrying hard, stringy meat. The Sussex to-day is quite as quick-growing as any of the beef-breeds, is deep, and, when thoroughly well finished, the carcase carries a very fine quality of meat. It is, moreover, famous for its hardiness, more especially as regards the breeding stock. Not only will the herd of cows stand exposure on cold clay lands located in exposed situations that might well be expected to kill cattle of ordinary constitution, but they will thrive under these trying circumstances on food that would be considered altogether inadequate for stock by those accustomed to other breeds. I have known a breeder leave the thistles on his farm uncut so that the Sussex cows might feed on them in time of drought. I have seen the herd of cows belonging to this farmer grazing the thistles, two feet high, with great gusto and thriving on the weed almost as well as if they had been browsing decent pasture-grasses.

It cannot, on the other hand, be denied that the Sussex inherits some of the faults of the draught-cattle from which he springs. He has not the symmetry of many breeds; he is very liable to be coarse; and the hard muscle of his progenitors has left undesirable traces of its presence in the animal economy of some stock of the present day. Admittedly the Sussex bullock has to be thoroughly fattened before he is a really good butcher's animal. If the beast is slaughtered before he is thoroughly finished the meat is tough, unpalatable, and of a bad colour. It is only when the bullocks are thoroughly prime that they yield carcases of the firm meat that attracts the butcher's customers.

The great hardiness of constitution, upon which all who

know the breed are agreed, is so valuable an attribute that it is worthy of investigation. No one, as far as I know, has ever investigated the factors which allow the Sussex to thrive under privations which might be fatal to cows of other breeds. This strength of constitution, however, has not of itself sufficed to win the breed any general popularity. Those who keep Sussex cattle are enthusiastic in their praise, but they are few. In Sussex and the surrounding counties other breeds are much more numerous. Neglect of the breed would be intelligible if it were ignored because of its want of milking power, but this is not so. There is a vast area of good grazing land on Pevensey Marsh that was till 1914 almost entirely devoted to summer grazing for beef; but the stock to be found on it were not of the Sussex, or local, breed; yet the adjacent Romney Marsh, devoted to sheep, was populated almost exclusively by flocks of the native variety. The absence of Sussex cattle is all the more inexplicable when it is remembered that there are a great number of farms on other geological formations in that neighbourhood which before the war were under permanent grass only good enough for breeding store-stock. The exposed position and damp nature of the land of these farms would, one would have expected, have made them specially suitable for producing the local breed. Yet with Pevensey at their doors the farmers who bred Sussex were the rare exception and the store cattle on the rich pastures came as far afield as from Herefordshire, Wales, and other equally distant places. When one did see the exceptional sight of a bunch of Sussex on these rich grass tracts they seemed to be doing quite as well as their neighbours; I never heard a complaint of them from those who had tried them, and the whole position is still a mystery.

The Sussex in one respect are especially good graziers. Amongst most cattle, turned out to get fat on pasture, the keen observer may see a good deal of picking and choosing. The beasts will frequently lift their heads and move on till they find a portion of the grass that particularly pleases their palate. If considerable care is not taken, the field will soon become rough and irregular in surface through patches of the coarse grasses being left untouched and the shorter and sweeter herbage being bitten off too close. The Sussex cattle, on the other hand, feed straight across the field, eating the pasture as it comes; slightly less land should therefore be needed to fatten them, and a better covering of grass can be kept on the pasture they are grazing. Of the commercial value of this peculiarity I have no idea. The general statement "one acre of finishing land ought to do a bullock" is apparently considered to be sufficient information for the present-day husbandman. It is approximately true that the average best pasture will graze an average large bullock, say, weighing 1,000 lb., and that in the grazing season of 20 weeks one may expect, taking the average of seasons, that each acre of land will supply food enough to the animal to allow of its becoming fat, and that while grazing for this length of time on such land its weight will increase (again on the average taken over a great number of animals) by 20 imperial stone, or 2½ cwt. It is true that there are many difficulties in the way of obtaining more definite information. The greatest of these difficulties is that of expense. Up till now it has not been found worth while to spend money upon such research. But there is so much variation among individual bullocks that it seems foolish in the extreme not to obtain more exact figures. In the very limited data I have been able to compile in respect of animals receiving the same ration, I have found a truly wonderful difference. One is therefore justified in stating that it would be worth while to find out whether or no the Sussex is a more economical grazier than an animal of other breeds.

Great credit is claimed for the Sussex bull as a begetter of good beef animals from thin-fleshed milking cows[1]. I have certainly seen very many specimens which, judged by the eye, gave much support to this contention; but, on the other hand, the demand for the services of these bulls is very restricted, and during my 20 years of observation the breed has not seemed to be on the increase. The extraordinary hardiness of these cattle seems to me to be the factor of greatest potential worth to the future. Improvement of beef-production, coupled with increased corn-growing on cold clay land farms in exposed situations, should, however, be a very important and valuable

[1] See "Sussex Cattle" by H. Rigden, *R.A.S.E. Journal* for 1908.

national asset to our future agriculture; for no one can deny that now there are many districts where little meat is produced and no corn at all. And it is to be remembered that it is frequently under these very circumstances that cattle can, by means of the farmyard manure they manufacture, make it possible to grow profitable crops of corn. Further, it is quite conceivable that, with more knowledge of the science and practice of breeding, the constitution of the Sussex might be combined with some good qualities he does not now possess. I have seen some particularly good beeflings of this breed but always among those which had been forced with very high feeding; I have great doubts about their being made good enough on mere commercial rations.

The South Devon.

It is a great pity that the old local name of "South Hams" has not been adopted by modern breeders, for this very large, strong-boned, pale-red, dual-purpose animal. We have, as has previously been observed, a subdivision already existing among the small neat *Devons*. My own journeys to study the South Hams have led me at least as much *into Cornwall* as into Devonshire! No two breeds could be more dissimilar; nevertheless, following the tradition by which the nomenclature of farming has been persistently confused, the Herd Book Society determined, some 30 years ago, that "South Devon" should be the name of the cattle which they formed themselves into a corporation to improve.

There are several interesting points about this huge, and it must be admitted, ungainly beast. Its size is very great; its skeletal development is enormous; among the cows are to be found very deep-milkers indeed, though they vary very much in this respect, and they have, as one might expect, tremendous appetites. The milk they give has locally the reputation of being specially fine for purposes of making clotted cream, and the beautiful butter that originates from the local production. This is what the outside world knows as "Devonshire Cream," but west of Plymouth such a description will be very emphatically corrected by the assertion that Cornish cream is better, but without any explanation as to the difference between the Cornish and the Devonshire article. The reason why this strong-boned, flat-sided cow's milk should be held to be so good for clotted cream seems to be rather a matter of colour, as the South Devon's milk is particularly rich-looking. There may be other reasons, but so far they are unexplained. Aged steers of the breed are much sought after to graze the rankest of the best pasture fields in the Midlands, and there is a very successful public sale at Totnes every year for the distribution of the bulls which are by no means purchased only by local farmers.

The astonishing thing is that meat-salesmen do not object to South Devons; by a certain class of butchers they are very highly valued. Though the well-bred bullock cannot be accused of being coarse, there is such development of bone that one would expect this seriously to interfere with his popularity on the block; moreover, for his size, the proportion of first-quality joints always appears to be small both to the eye and under the hand, and the "roastings" seem to be thin-fleshed; generally, to the uninitiated, the South Devon seems very unlikely to attract the butchers' favour. I have often been told, by those experienced in the trade, the cause of this unexpected popularity. The beast is found to be particularly thick-fleshed at those parts of the carcase **which carry the second-class cuts.** These are parts which are often difficult to sell. The tradesman who has to supply, often under contract, large joints off the first-quality parts of the carcase, often finds himself with a quantity of meat which is cut from other places that cannot easily be disposed of. My informants are convinced that the second-quality cuts from the South Devon are so particularly good that it is comparatively easy to find customers who will pay a profitable price for them. Certainly, I have noticed that these steers when on the hoof seem to handle particularly well at the brisket, which is one of the largest, and a very important, joint of second quality. It would be interesting to know whether the South Devon is exceptionally well covered with meat at such points or whether he is merely better covered there than might be expected from his general conformation. For, if these cattle should prove to have a superiority over all others in this respect, they might, with their possibilities for milk-production, be of use in future breeding for general improvement.

Welsh Black Cattle.

Some 20 years ago there were two distinct types of black cattle in Wales; the North Welsh and the South Welsh, each with its own Herd Book. For some reason the Breed Societies of the two varieties amalgamated in 1904, and we now have but one Herd Book.

The marked difference of the two sorts is still to be seen in the markets, the Southern type being stronger in the bone, less well sprung in the rib, and longer in the leg. The horn of this type is often very dark indeed, and this was frequently looked upon as a danger-signal when buying store cattle to grow on into beef. The South Welsh cow was regarded as the better milker and was held in high respect for the constitution which enabled her to fill a moderate-sized pail under most adverse conditions of climate and keep; but when it came to buying the steers to which she had given birth, admiration ceased. Slow-growing and thin-fleshed, these steers would live on barley-straw in the open throughout a hard winter, and finish on moderate grass in the spring or summer at the age of three or four years; it was very ungrateful work trying to force them to better things on superior food. They had a strong objection to any kind of close confinement in the winter, and did better in a large, open yard with a little hay and plenty of straw than when confined in a stall on the most abundant and nourishing ration. Their chief virtue was their ability to stand the most trying conditions of climate combined with the privations of rough forage.

The North Wales cattle, though the cows were not considered to milk as well, bred the famous store cattle known as Runts. These were held in very high esteem by all graziers feeding stock for beef on good pasture; and especially by those wanting cattle to graze the fertile but exposed marsh pasture-land on the east coast of England. The Runt was a moderately well-shaped butcher's beast, a little coarse at the rumps, not particularly well sprung on the rib and, except in the best bred specimens, just a little long on the leg. On the other hand, butchers testified to the abundance of lean meat which they carried. They had, further, the reputation of yielding a very high percentage of carcase to live-weight, and though I have never seen any beeflings of this breed, the Black Welsh cattle at the Smithfield Shows gave quite good weights for age.

The Agricultural Commissioner for Wales, writing[1] of the cattle of the Principality, says, "There are breeds that grow and feed more rapidly than the Welsh; there are breeds that are better known for their milk; but there is none which combines general utility with the hardiness necessary to thrive under very unfavourable conditions to a greater extent than this breed." If the South Wales steers such as I tried to winter-feed in my youth are excepted, and if the milking qualities of the dams of the Runts are shown by their records to be good, no one can say that this claim is unfounded. If, on the other hand, national ardour has led to some bad-doing steers and some few doubtful milkers being overlooked, there can still be no doubt that the systematic or scientific breeder has in the Black Welsh cattle very promising material for work in the future. The contour of the land in many of the beautiful districts of Wales is such as to ensure its being used rather for pastoral, than for arable, husbandry. Thus the improvement of an already hardy and useful breed should be full of interest and promises a rich reward.

[1] See *The Standard Encyclopedia of Modern Agriculture*, vol. 12, page 125. The Gresham Publishing Co. Ltd.

CHAPTER XII

POLLED BREEDS

The Aberdeen Angus.

As the name indicates, these cattle originated in the herds on the farms of the north-east of Scotland. They are normally polled, though very occasionally horned animals appear among them; small, immature weapons of defence, called *scurs* are, however, fairly common. The prevailing colour is a jet black; occasionally white markings appear and whole reds are by no means unknown, but the vast majority of breeders aim at producing stock of the first-named colour only. When thoroughly finished the breed is, as a butcher's beast, very near perfection. As it is all to the good for domesticated cattle to be without horns, it is surprising that the breed does not spread itself over the whole of the world's surface where cattle are required to range and grow fat.

The general appearance of the high-bred bullock is blocky, square, and very deep; the whole animal exemplifies quality, and is especially good at all the first-quality joints. Along his back the ribs and loin are very wide and carry a large amount of well-marbled meat; I have even heard butchers complain of their being too good at the "roastings," as these parts are called by the trade; the reason given for this strange complaint was that customers, being usually obliged to rely on cuts from other stock, were spoilt when "Aberdeen Angus joints" were supplied for a short time!

Furthermore, the beast is very level-fleshed; that is to say, there is an absence of all that patchiness about the body which denotes excess of soft fat; he is also a wonderful handler, being firm and springy all over. "A mass of firm meat" is a frequent comment after a manual examination, and well describes the quality of the animal.

The fleshers contend that not only are the "roastings" and other first-quality joints excellent, but also that the flavour and texture of all the meat are especially fine. During his first year the Aberdeen Angus is perhaps not very fast-growing, but later he fully makes up any deficiency in size; he is not a particularly large feeder, but is credited with great powers of digesting coarse or inferior fodder. On the famous north-country turnips and oat-straw cattle of this breed are reputed to thrive and grow amazingly. When mated with good specimens of almost any other beef-breeds, superb beef-making cross-breds are the result.

Yet with all these points to their credit, Aberdeen Angus bulls do not monopolize the market when buyers want sires to improve, or to maintain, the beefing qualities of great herds of cows; this is especially the case when the troops of females are inferior, or unimproved, stock. What makes this the more surprising is that specimens of the breed have carried off the Championship at innumerable public exhibitions all over the world in competition against animals of other breeds. Perhaps the explanation of the Aberdeen Angus bull not being in greater demand is that he has not the power of stamping quality and quick growth on his offspring when mated with cattle inclined to be thin-fleshed and slow-growing. This conjecture is supported by the evidence, frequently to be seen, of bunches of very inferior store cattle evidently begotten by bulls of the breed. Such evidence, however, is liable to be misleading for the following reasons: the best first-cross stores are so much appreciated that they very frequently pass from one owner to another by private sale, and the absence of such rarities of the trade from the public markets may make the misfits all the more prominent to the eye. The strange thing about this failing, if there be such a failing, is that the breed itself has a very great propensity for piling on fat. Aberdeen Angus cows or heifers more often fail to breed through getting excessively fat than any other breed I know; at any rate this is the cause given for their failure, and certainly appearances often confirm such a view. That there are good milkers among cows of the breed is undoubted, and on the whole one is driven to the conclusion that the factors which go to make up this wonderful breed ought to be thoroughly investigated with a view to making the Aberdeen Angus of more general use to the tenant-farmers of England.

The cattle traders of Scotland who specialize in the very high-class meat that the breed undoubtedly produces have a very fine system of marketing. Either on hoof, or as carcase, the cattle are distributed methodically so as to prevent an excess of supply over demand lowering the value of the best quality. The auctioneers in pre-war days, at markets on the east coast of England, would receive just enough to satisfy the wants of the best tradesmen in the seaside-resorts; large consignments would go to large centres of consumption; small ones all over England where wanted; even Paris received its proper amount of the famous Aberdonian "Roast-Beef." It is most earnestly to be wished that such systematic marketing may soon become universal in Great Britain; if the producer and consumer are to have a greater share in the produce of the soil, there is indeed great need for such co-operation.

The Galloway.

This breed, like the last, is polled and most frequently black in colour, and for these reasons the two varieties are often mistaken for one another. It is true that underbred specimens of each variety are apt to be very much alike, for they lose the breed type of both; but, apart from this, it may be said that the two breeds differ as much as the south-west of Scotland— the place of origin of the Galloway—and the home of the Aberdeen Angus in the north-east of that country. As they give a good opportunity to the student to exercise his judgment, it may be useful to describe the differences.

The head of the Aberdeen Angus is shorter, especially from eye to muzzle; the lower jaw is deeper, better sprung and cleaner cut, and firm at the muzzle. The width between the eyes is great in both breeds. The short, wide head with well-sprung, deep, clean-cut jaw is often looked upon as typical of good meat-production quality. It is certainly true that this is the shape of the head of the Aberdeen Angus and it may be admitted that that breed is second to none as a good beef-producing animal; but it is quite conceivable that the head of the good beef animals that are not Aberdeen Angus may be of other conformation. It may well be, indeed, that the Aberdeen's head is the origin of the popular belief that a certain shape denotes beef-producing qualities.

The poll of the Galloway is flat, that of the Aberdeen Angus is pointed, and the ear of the former is larger and less well-rolled, or more open than that of the latter. The hide of the Galloway is much the thicker, and very highly valued by the tanner; its coat is longer and more hairy. The greatest difference however, is in the shape and build of the whole body. The Aberdeen, well sprung in the rib, thick in the flesh, fine in the bone, and well put together, is well described as "block-like"; the Galloway on the other hand is quite strong-boned, has rather flat sides, looks longer, and is what a horseman would describe as "loosely coupled." Thus the two breeds are quite different in shape.

The colour of the Galloway often shows a red tinge; black and whites, duns and reds are also frequently found. The breed has a great reputation for hardiness against which must be set the charge, made by those who know the breed that they do not grow quickly or mature early; it is, as beef-breeds go, a good milker, but does not generally give a high enough yield to claim dual-purpose capacity. It is sometimes claimed for it, and perhaps with some justice, that a joint from its carcase will give a deeper layer of lean meat, in proportion to its size, than that from any other British ox.

It is, however, the power of the cow, when mated with a "white" Shorthorn, to drop the wonderful "Blue-grey" calves which make the highly valued stores, that renders the breed most interesting. Bulls of the desired colour have been bred for this purpose in the "Border Country" for generations by many farmers who make a speciality of supplying the demand for these sires. I have wondered whether these bulls are truly described as white; to me they have always appeared to be *cream colour*. It is a point of some interest, having regard to the constant discussion of the inheritance of colour according to the laws of Mendel. The cross is altogether an admirable one as regards beef-production. The quality and early maturity of the Shorthorn parent prevail, and to this is added the firm fleshing and, according to some authorities, the abundance of lean meat from the Galloway dam. Blue-greys are said to be more hardy than Shorthorns, but I have no experience in this, and probably the point is only rarely put to the test, for, after weaning, the calves are well cared for; it is the cows that are expected to stand a great deal of exposure in the altitudes of the rigorous climate in which they live. Nor have I ever been able to find reliable evidence that Blue-greys "do" particularly well; this term, implying a profitable return for food consumed, is, however, frequently applied to them by men of experience. One is compelled to distrust popular beliefs founded upon appearances. When neither the animals nor the rations are weighed, obviously the eye is the only guide. How often the eye is misleading only those who have experience of recording weights and amounts of food consumed can thoroughly realize. The beautiful Blue-grey always shows to advantage; it is very seldom that a beast of this colour looks otherwise than thriving, and so it is quite possible that, when judging by appearances rather than by ascertained fact, this colour creates a prejudice. The belief is so widely held and the knowing feeders pay such high prices for the stores, that it is important to learn the truth. Until we know the truth, it is difficult to dogmatize about Blue-greys. The value of first-cross bred animals is held to be very great. Many practitioners will assert freely that the first-cross produce of two breeds is better than either of the parents. The origin of this belief can very frequently be traced to the alleged superiority of the Blue-grey; and to accept the whole theory on this assumed superiority seems most unwise. The importance, then, of testing the belief by proper trial is clear.

In plant-breeding there appears to be greatly increased vigour in the "hybrid" offspring of two different varieties crossed together. Many observers working on Mendelian lines have noted this. It may be that the Blue-grey is an instance of what Mendel's disciples call an F_1, showing the same increased vitality as is believed to be seen in the plant. On the other hand, it is permissible to doubt whether the factors governing utility characters are fixed to the same extent in the different breeds of cattle as they are in different varieties of plants. It has often seemed to me that, as regards utility points or characters or aptitudes, F_1 animals are apt to occur within the same breed. This possibility might, with care, be tested while ascertaining the relative qualities of the Galloway, the Shorthorn and their cross-bred offspring—the Blue-grey.

The Red Poll.

The only remaining polled breed that is bred in sufficient numbers to be looked upon as commercial stock is included in this chapter because of its hornless condition, and not because it has any resemblance to the two polled breeds of Scotland. Nor is it intended to countenance the vague speculations on the polled condition now found in the East Anglian breed—a breed which to-day differs in many important respects from the Northern cattle. Indeed, even this remote and ancient connexion which is held to have had such important results as regards the carrying of horns is not admitted by all authorities, and it is therefore safer to say that the red cattle of Suffolk and Norfolk simply resemble the Aberdeen Angus and the Galloway in being hornless and that there the resemblance begins and ends.

Some years ago, not many generations, as cattle lineage goes, it would have been true to say that the Red Poll was *par excellence* a dual-purpose breed. That one cannot say so now about all the tribes or families composing the breed is due to certain distinguished breeders having determined to select their bulls and cows with a view to beef-production to the neglect of milking qualities. It is very difficult to understand what gain was to be expected from this policy. There are surely many breeds as good for the shambles as any Red Polls that have ever been produced; there is no evidence that the best specimens of Red Polls were more economical in any way as beef-makers than the general run, say, of the Aberdeen Angus or the Devons; and the former have never rivalled the outstanding stock from the herds of the many recognized single-purpose beef varieties. On the other hand, the dual-purpose animal is much less common, is much more useful in intensive husbandry, such as ought to be commonly practised in a thickly-populated country, and the double faculty of meat and milk production is very easy indeed to lose. The Red Poll breeders may well be reminded how easy it is to lose milking qualities without necessarily improving the fleshing capability of their stock. The Shorthorn breeders' experience ought, it is suggested, to act as a wholesome warning to the owners who still are fortunate enough to own herds of dual-purpose Red Polls. It was a common experience fifteen years ago, it is still unfortunately far from rare, to breed a Shorthorn that gives the smallest possible amount of milk and yet cannot be classed higher than as a second-class beef-producer. The live-stock literature of the seventies is full of the surest evidence that certain Shorthorn families were then excellent milkers. Unfortunately these families were in some cases mated, through several generations, with sires that had been bred solely for beef, and so the cows of the tribes which in the seventies had been famous for the dairy had lost class as milkers in the nineties. At the beginning of this century, or thereabouts, when a determined and successful effort was made to re-establish the dual-purpose Shorthorn in all her glory, the folly of the previous method of breeding became only too apparent.

Within the last ten years I have bought for grazing, in a good midland farming district, a truck-load of pedigree heifers at less cost per head than I could have bought non-pedigree steer stores from Ireland. There was not a heifer in the whole bunch whose pedigree was not good. The volumes of the *Live Stock Journal* and *Fancier's Gazette* for the years 1875 to 1880 bear testimony again and again to the good dual-purpose quality of cows that were the progenitors of these heifers which, through the use of pure beef bulls, had lost all aptitude to milk and had not thereby gained one scrap of flesh. This was no uncommon experience. There are to-day very few Dairy Shorthorn herds in which, after years of careful selection, an occasional specimen, such as I have just described, does not appear. The Red Poll breeders will, therefore, be well advised to preserve the dual-purpose character of their stock with the greatest possible care and to guard against using bulls, however blocky and full of flesh they may be, whose female progenitors are not known to be good milkers. Indeed, the milking qualities should be guaranteed by the dams and grand-dams of all stock bulls being published in duly authorized milk-recording societies.

The Red Poll is of medium size and long in the body; and though well sprung in the rib, is not particularly deep even for her size. For a medium-sized animal, the milk-records from the carefully-bred herds are very good indeed. The milk is of good colour, and decidedly rich in butter-fat, though, as in all breeds, quality varies greatly in individual animals. In beef-production the breed has many superiors, and butchers are somewhat critical of the steers. Still, considering the moderate rations required for a beast of this size, the Red Poll steer is infinitely superior to the majority of the mongrels which the unhappy plough-land farmers have now to feed. The colour, though always red, varies considerably in shade. It would appear that the good milking strains tend to be of the brick-red, and the beef strains of the deep, blood-red tints; this may, of course, be accidental, but there are authorities who hold that the yellow, or brick-red, colour is an indication of deep-milking proclivity.

Besides the hornlessness and her dual capacity, the cow of this breed can be relied upon to drop a male calf that will make an excellent beefling. The flourishing cattle-market at Ipswich, in the very heart of the Red Poll district, has a great reputation for these small butcher's beasts, and among the specimens of such sold at this centre few, if any, are preferred to the native cattle.

I have heard many traders at this, and other neighbouring markets, speak with little enthusiasm about the aged steers of the indigenous breed, but I have never heard anything but praise of those turned out as baby-beef. The combination of good qualities, together with its medium size, makes the Red Poll especially valuable to those who wish to farm intensively on light or medium land. For the supply of beef, with corn, to the national larder from land that will not grow heavy crops it may be asserted that the breed probably has no superior.

CHAPTER XIII

BREEDS (*continued*)

The Dutch or British Friesian.

The black and white cattle of Holland have acquired great notoriety during the last four years owing to the truly amazing prices made by them in the auction ring. These fabulous prices culminated in no less than 4,500 guineas being given in the summer of 1918 for a specimen of the breed. Less than ten years ago I spent a fortnight in Holland buying these cattle, and throughout my stay met many prominent breeders who frequently alluded with pride, as well as with evident surprise, to the fact of the champion bull being sold for £400. It is true that £6,000 has in the past been paid for a pedigree Shorthorn cow and that high four-figure bids were constantly made for specimens of the breed in the early seventies of last century. But this period of what has been called the Bates and Booth mania occurred no less than 60 years after the whole agricultural world was set talking about the 1,000 guineas paid by a combine of breeders for the bull *Comet* at Mr C. Collings's sale in 1810. Further, since the era just alluded to, £3,000 is known to have been paid for a champion at our leading live-stock show, and over £1,000 has not infrequently been given for outstanding winners in Shorthorn classes and prize-taking cattle of other breeds. In the prices recently paid for Dutch cattle, however, not only has the rise been extraordinarily rapid, but also the curve has mounted out of all proportion during the last two years, during which there have been no exhibitions to enable the stock to intoxicate the bidders by their triumphs in the show-ring. The facts are so astounding that they require some examination.

There can be no doubt that considerable prestige was given to the breed when, in the summer of 1914, the Government permitted some of the stock to be imported from Holland, notwithstanding the statute which excludes all cattle, except those very specially exempted from its adamant regulations. So strict and rigid are the regulations closing our ports against the entry of any cattle for breeding purposes that, when it was first whispered that a permit allowing importation had been given to the Association formed to advance the interests of this breed, the news was discredited; but the report proved true. The British Holstein-Friesian Cattle Society had purchased 39 bull calves and 20 heifers in Holland during the month of July 1914. Under the strictest surveillance a long period of quarantine was enforced, and then the cattle were offered at an auction sale at which the average price was £253 per head, the highest figure made being £560, the lowest £110. The profits on the whole transaction, which amounted to a very considerable figure[1], became the property of the Society, and full credit can be given to this body for making good use of the windfall in forwarding the interests of their breed. Further, the action of the Government in making this great exception to the regulations aroused much interest in the minds of all concerned in the cattle industry and added not a little kudos to such merits as the breed possesses. These and doubtless other matters, such as currency, the scarcity of milk, and the known performances of the best specimens of the breed in their own and other countries, led to prices at sales rising till the transactions resembled those of the gaming-table or racecourse.

It therefore behoves the farming population to examine the intrinsic merits of the breed rather than to seek guidance from the accounts of results obtained at public auction. To be in a position to judge the merits of the breed one must thoroughly realize the difference between those kept during the last 30 years in Holland, a few of which were imported in 1914, and those

[1] The *Farmer and Stockbreeder* of November 9th, 1914, gives a full report of the sale, showing the prices paid for the cattle in Holland and those, obtained at Byfleet in Surrey, when the cattle on coming out of quarantine were sold at a semi-public auction arranged by the British Holstein Cattle Society. This well-known periodical makes itself responsible for the following statement "that there is a little over £10,000 clear profit which goes to the British Holstein Society." The 59 head of cattle, mostly calves, cost less than £5,000 in South Holland and sold for £14,936. 5s. od. One of the young heifers sold that day made the 4,500 guineas alluded to above; the highest price made by an imported female at the Byfleet sale was £520!

kept in Britain during that time. The indigenous cattle of Holland were wanted to milk; with this object in view, careful and well-authenticated records of performances at the pail were kept, so that all cows giving small yields could be eliminated—and this was done ruthlessly. Furthermore, the milk was more often than not sold at rates which rose or fell according to its richness in butter-fat. This was necessary because this produce was wanted for cheese and butter-making, and also because it was found that the unimproved breed naturally gave milk of low quality. By carrying on a system of milk-testing for quality while recording the quantities, the breeders of Dutch Cattle, in the course of years, threw out from their breeding stock all parents likely to reproduce bad milkers, either in the sense of giving little or of yielding stuff of low fat percentage. In this way the Province of Friesland and the greater part of the rest of Holland obtained, after many years of good work, cows that could be relied upon to give great weights of milk of moderately rich quality. More than this cannot be said; for the average quality of the best herds in Holland is not high in butter-fat.

Now in England no effort was systematically made under supervision, as was the case in Holland, to get rid of bad milkers up to the year 1914; the result is that, whenever it has been tested, the milk varies very much from moderate to poor in quality. In fact, the fluid is often so poor as to contain less than the 3 per cent. of butter-fat which is the limit fixed by Government as a standard below which milk may not be sold; or rather, the authorities say that if milk is sold below standard it is, as has been said before, *held to be adulterated until the contrary is proved.* With all deep-milking breeds, untoward circumstances will occasionally bring the quality down to a percentage much below normal   For instance, if the average quality of the mixed milk of 20 cows is 3·7 per cent., one may reasonably expect that it will nearly always pass the standard; the exceptions, in other words, will be so few as to make the chances of a summons for assumed adulteration uncommon. But the unimproved Friesland cows, that is to say, animals that have not been bred through several generations from animals selected as breeding stock because their produce has been found to be relatively rich, may only give milk averaging from 3 to 3·2 per cent., and this quality is fairly certain, in the course of time, to bring the owner more than once before the magistrates. Thus, besides frequently supplying the public with inferior goods, the owner of unimproved Dutch cattle is likely to find his legal expenditure robbing him of much profit, even when he has a deep-milking herd.

This last-mentioned characteristic cannot be assumed to belong to all the cows now spoken of as "British Friesland" when the magic word "imported" is not prefixed to their name. On the contrary, the Black and White Dutch Cattle bred in England for several generations without any admixture of imported blood—an impossibility between 1894 and 1914—and without any controlled or systematic selection of breeding stock, are apt to vary just as much as any lot of milk-like looking cows, whether registered pedigree or mongrel stock. They are found to give, when in full profit, anything from 550 to 1,000 gallons of milk in the 10 months following the third and subsequent calvings.

The difficulties of establishing good milking qualities among these cattle are not so great as to deter our breeders from the attempt. Only an ignorant person could believe that the Englishmen who have won the admiration of the whole world as exhibition live-stock producers are likely to fail if they turn their attention to the production of good milkers. It may be safely asserted that the British breeders will, if they so wish, in the course of a few generations place their own Friesian cattle in the highest possible class for quality and quantity of milk; and we know that in regard to milk-production the stock belonging to the best Dutch breeders is capable of giving very high yields of moderate quality. In this respect our home breeders have a very severe task in front of them, and they know it; but they may be trusted to be large-minded enough to profit by the wisdom of others and to use the same scientific methods which their competitors have used for some 30 years with such successful results.

Granted that the necessary skill can undoubtedly be found to accomplish all this, one is still bound to wonder whether this breed is suitable for our own husbandry. It is much to be regretted that our Government did not establish this fact before encouraging, by allowing an exceptional privilege to the breeders of these cattle, the reproduction of the race. There can be no doubt that a craze (one can use no other term) has set in for the British Friesland which may possibly be detrimental to the future production of beef in our home country. The whole question turns upon whether or no the cows can breed beef steers. They are exceptional milkers as regards quantity. They are far behind Guernseys as regards quality of milk, and the Guernsey could easily be selected to compete with the Friesland in regard to quantity if size were taken into consideration; thus there is no reason, from the milk-yielding point of view, why we should find the former breed selling for hundreds while people tumble over one another to bid thousands for good Friesians. The Dutch cow, if slaughtered young, yields a carcase of good cow-beef, but no better than is to be obtained from any good non-pedigree deep-milking Shorthorn. Of these latter it may be objected that they vary very much, but in the districts of South Holland—a locality from which the best dual-purpose cattle of Holland are to be obtained—- I have never seen cows anything like as good as the *best* Shorthorns sent out regularly every month by the dairymen who supply the North Country cities of our own land with warm milk. The latter have a just grievance against the breeders of the worst cattle they have to house and fatten for the butcher; they have a right to ask that the English breeder should emulate the wonderful uniformity of the Dutch breeders' stock; but they have always among their herds *some* deep-milking cows that for beef-production are quite unequalled by any dual-purpose cattle in the world. The Friesian drops a very wonderful calf, considered from the point of view of feeding for veal. Though I have no means of judging other than by observation, nevertheless, I feel convinced that no breed produces a calf that is born so fat as the Friesian; the importance of this factor can only be realized by those who have been obliged, to their misfortune, to get veal from calves which, at birth, were mere skin and bone. But the whole problem before us is to maintain our reputation as prime beef-producers and, while doing so, to make a reputation in the future for farming intensively the whole of our land, not merely the more favoured areas, thus winning abundant food for the nation, besides profit for the individual. For this purpose cattle that are good milkers, even though they drop calves that make good veal, cannot be said to be suitable if they do not breed steers suitable for beef making.

I am in considerable doubt as to whether the Friesian has this latter quality. For many years I have made observations on this point. In the Netherlands I have never seen one good steer of the desired kind. It is true that one does not see many, but all I have seen have been inferior. In this country, every now and then, one hears of good Dutch beef-beasts, though far more frequently one hears complaints about them. I have never seen one, that I knew to be pure bred, of better quality than those I have seen either in Holland or those brought from that country to the rich cattle pastures of Flanders. The good bullocks shown to me on more than one occasion were, obviously, merely Dutch in colour and markings, and many times I have been able to establish the history of cross-breeding. Black and White animals, with the typical Friesian admixture of the colours, are very persistent in cross-bred cattle that carry any Dutch blood in their ancestry. It is such cattle as these, I am inclined to suspect, that are the source of the rumours one constantly hears of the good fleshing qualities of so-called Dutch beef-beasts.

In the preface that Dr F. H. A. Marshall has kindly written for this book, he speaks emphatically (see p. vi) of the complicated anatomical characters that go to make a good beef-beast. There is a possibility that the Dutch steer may fail in one or more of these characters, and it is most important that this matter should be speedily put to the test of thorough and impartial investigation. In the interests of the breed itself it is desirable; for there have been so few real specimens available for empirical observation that it may be found to be quite wrong to deny good beef-making qualities to the breed. On the other hand, claims are being put forward that the high prices paid at public sales for specimens of the breed are justified on account of the animals possessing dual-purpose capacity. If by dual-purpose capacity it is meant to imply the power of breeding steers fit for the feeder and butcher, it can only be replied that the claim is made without any sound evidence to support it. It was rumoured at the time of the famous sale of imported stock in 1914 that the Government had made this most exceptional concession in favour of the breeders of this variety for the sake of the cattle export trade; if this was so, well and good. I happened to be in South Holland, North Holland and Friesland at the same time as commissioners from America, Russia, Japan and South Africa who were buying these cattle, and I never heard the dual capacity of the breed mentioned. All reference to the animal's merits was confined to the records of milk-yield and butter-fat tests held by various females in their pedigree. I am, therefore, not a little apprehensive that the merits of the breed as beef-producers have been urged with a view to securing the custom of our home tenant-farmers. If this apprehension is well founded, it is very deplorable; for, if we are to avoid the evils of the past, the home husbandry should be our first claim.

No one will question the value of the export trade. But if our future husbandry is to be as productive as possible, the subordination of the tenant-farmer's interests to the demands of the overseas trade must be carefully avoided. The greatest possible credit is due to the body of pedigree-breeders who have captured and held the export business, for the animals they have sent all over the world have deservedly won unstinted admiration; but the demand they have thus supplied is not necessarily the same as that which should come from their neighbours. In the past, the home supply of suitable stock has been inadequate, while the customers from overseas have been fully and admirably satisfied. There is no need whatever to curtail the efforts of those who wish to continue the overseas trade; there is, in fact, every reason to encourage them, but the home supply must, if we are to maintain ourselves in the forefront of nations, be for the future the first consideration of those responsible for the advancement and improvement of our breeds of cattle.

Till it is established, one way or the other, that the Friesian is suitable for the home requirements of our farmers, no public authority should advance the interests of the breed against those that are found to be useful. Until the question of beef-production among the steers is decided, it may well be that the enthusiasm for the breed which is being encouraged by many well-wishers to British farming, is but another case of failure to obtain the greatest possible amount of home-grown produce from our land. Here again is seen the great need for more of that knowledge which can only be acquired through scientific research and systematic trial, tested by practice on the farms of our native land.

INDEX

Printed in the United States
By Bookmasters